Polymer degradation and stabilisation

POLYMER DEGRADATION & STABILISATION

NORMAN GRASSIE

Professor of Chemistry, University of Glasgow

GERALD SCOTT

Professor of Polymer Science, University of Aston in Birmingham

CAMBRIDGE UNIVERSITY PRESS

Cambridge

London   New York   New Rochelle

Melbourne   Sydney

Published by the Press Syndicate of the University of Cambridge
The Pitt Building, Trumpington Street, Cambridge CB2 1RP
32 East 57th Street, New York, NY 10022, USA
10 Stamford Road, Oakleigh, Melbourne 3166, Australia

First published 1985

Printed in Great Britain at the University Press, Cambridge

Filmset by Mid-County Press, London SW15

Library of Congress catalogue card number: 84-9617

British Library cataloguing in publication data

Grassie, Norman
Polymer degradation and stabilisation.
1. Polymers and polymerisation – Deterioration
I. Title   II. Scott, Gerald
668.9   QD381.8

ISBN 0 521 24961 9

MP

CONTENTS

v

PREFACE

Organic polymers, by their nature, are much more prone to chemical reaction than most traditional materials. The use of polymeric materials in increasingly demanding applications has led in recent years to an upsurge in the study of polymer durability both in industrial laboratories and in universities and research institutes. The importance of the subject is reflected in the appearance of a new journal and several specialised series of publications dealing with a broad range of degradation and stabilisation processes. From the practical point of view these studies embrace a variety of technologically important phenomena, ranging from low temperature processes such as the 'weathering' of polymers, 'fatigue' in tyres and the processing of polymers in shearing mixers, through to very high temperature phenomena such as flammability and ablative processes.

It is impossible in practice to deal adequately with these different technological problems in a single volume, and it is not the purpose of this book to attempt to do so. It is, however, fortunate that many of the technological phenomena have a common scientific basis in a limited number of chemical mechanisms which have been intensively studied in academic institutions over the past twenty years. Thus weathering has its roots in photo-oxidation, fatigue and melt degradation in mechano-oxidation, and flammability and ablation in pyrolysis and vapour-phase oxidation. This book will therefore attempt to relate technological phenomena to the chemistry and physics of degradation and stabilisation processes.

It is primarily intended for the third-year undergraduate, for postgraduate students specialising in polymer behaviour and for scientists and technologists in industrial laboratories who are concerned with the performance and particularly the durability of polymeric materials during manufacture and in service. This is the first time an attempt has been made to include all aspects of polymer degradation and stabilisation in an undergraduate text and inevitably the specialist will find it incomplete. In order to compensate for this, an unusually large number of review references are given at the end of each chapter.

We believe the work is justified on two grounds. First, it is important that graduates who are going to specialise in polymers should know something of this increasingly important subject before completing their degrees. Secondly, a grasp of the science underlying degradation and stabilisation processes should in the future provide a more effective pathway to the solution of diverse technological phenomena than the empirical approach which has been widely practised in the past.

N. Grassie

G. Scott

1

The practical significance of polymer degradation

1.1. Historical perspective

Man has been aware of the degradation of polymers from the beginning of his history as a manipulator of materials, although he would not have described it in these terms. The rotting of wood and cloth, the deterioration of meat and the burning of wood are all examples of irreversible and sometimes catastrophic changes which have constantly challenged man's control over his environment.

The earliest documented examples of the deterioration of modern polymeric materials followed the discovery of rubber by European explorers in the Amazonian forests. They were impressed by the physical behaviour of the product of *Hevea brasiliensis*, since its ability to rebound appeared to give it an independent 'life' of its own. On the long sea voyage back to Europe, however, this interesting behaviour was lost and since they believed that raw rubber as it came from the tree was in some way 'alive', they considered the rubber to have 'perished' during the journey. This kind of anthropomorphism has continued in rubber technology to the present day in the use of such terms as 'ageing' and 'fatigue' for different manifestations of the deterioration of rubber associated with its molecular weight decrease.

It was not until long after empirical solutions to the problem of rubber ageing had been found that the chemical reactions involved in the degradation of rubber were investigated and identified. Vulcanised rubber was sometimes found to 'age' better than raw rubber and this phenomenon was later shown to be due to the formation of by-products of the 'curing' reaction. These preservatives were identified as 'anti-oxidants', since as early as 1861 Hoffman had demonstrated that oxygen is involved in the degradation of rubber.

Subsequent studies have shown that other minor constituents of the atmosphere may also contribute to the deterioration of rubber. In particular, ozone, which is a contaminant in most industrial atmospheres, is a potent cause of crack formation of rubbers under stress.

The basis of the modern theory of autoxidation is to be found in the work of Farmer and Bolland and their co-workers at the British Rubber Producers Research Association (BRPRA). These workers applied Bodenstein's concept of a chain reaction involving highly-reactive intermediates (later identified as free radicals) to the process of autoxidation. Farmer showed the importance of hydroperoxides as the primary products of this reaction and most importantly, from the practical point of view, Bolland was able to show that certain readily oxidisable compounds in very low concentrations can inhibit autoxidation by interrupting the radical chain reaction. The modern theories of antioxidant action owe a great deal to the penetrating studies carried out at the BRPRA and its successor the NRPRA between 1940 and 1960. Bateman and Gee showed that hydroperoxides were also important photo-sensitisers for autoxidation. Surprisingly, this observation was subsequently overlooked by photochemists who championed the more strongly absorbing carbonyl compounds as photo-initiators for photo-oxidation. Recently, the hydroperoxides have been established as the most important photo-initiators in polymers during the early stages of photo-oxidation.

The discovery of the synthetic vinyl polymers led to a new problem in polymer degradation. It was early realised that the process of polymerisation is in some cases reversible, so that above the 'ceiling' temperature, complete degradation may occur to give the original monomer. In other cases, lower molecular weight products may be formed which are different from the starting monomer. Both phenomena seriously limit the use of vinyl polymers such as methyl methacrylate and polystyrene at high temperatures, and this has led to the studies, using a range of characterisation techniques, which will be discussed in the early chapters of this book, and later to the design of new polymer structures which are stable to much higher temperatures for specific applications.

Other polymers undergo a quite different process on heating in the absence of air. Poly(vinyl chloride) (PVC) provides the best-known example of a polymer which eliminates pendant groups along the polymer chain to give a highly unsaturated, and hence highly coloured, polymer without substantial chain scission.

$$-CH_2\overset{\displaystyle Cl}{\underset{\displaystyle |}{C}}HCH_2\overset{\displaystyle Cl}{\underset{\displaystyle |}{C}}H- \quad \xrightarrow{\Delta H} \quad -CH{=}CH{-}CH{=}CH{-} + 2HCl \quad (1.1)$$

It will be seen later (see chapter 2) that this occurs in an 'unzipping' reaction which is characteristic of a number of vinyl polymers with labile pendant groups.

1.2. Polymer durability

As we have already seen, polymers can be destroyed by a variety of environmental agents. In general, condensation polymers that contain functional groups in the polymer chain, notably polyesters, polyamides and polyurethanes, are much more subject to hydrolytic and biodegradation than polymers containing a carbon–carbon backbone. Indeed there is little evidence that the latter are susceptible to biodegradation at all unless they are first oxidised. This presents a problem in polymers required for short-term use and which are then discarded. The problem of disposing of plastics litter, for example from packaging, will be further discussed in section 1.5.

In the absence of light, most polymers are stable for very long periods at ambient temperatures. The effect of sunlight is to accelerate the rate of oxidation and this effect may be exacerbated by the presence of atmospheric pollutants capable of being activated to free radical species. This is particularly true of nitrogen and sulphur oxides, which are frequently components of industrial atmospheres. The more oxidisable polyolefins (e.g. polypropylene) are particularly sensitive to photo-oxidation, and the rate of physical deterioration of the polymer may be accelerated by almost an order of magnitude by light. This severely limits the use of polyolefins in the outdoor environment, and indeed, there was doubt for some years after the discovery of polypropylene as to whether it could even be developed as a commercial polymer. Some measure of the success of stabilisation technology is that polypropylene can now be considered for such outdoor applications as ropes and pleasure boats.

It is important to recognise that light alone is not an important degradative influence in the deterioration of most polymers. From both the practical and scientific points of view, photolysis and photo-oxidation of polymers must be clearly distinguished. They are quite different phenomena, and although the first step in the two processes is often the same, namely the photolytic formation of free radicals, the subsequent steps are quite different and have different activation energies and end-products. In general, photolysis leads primarily to unsaturation in the polymer, particularly at the broken chain ends, whereas photo-oxidation leads to the production of aldehydes, ketones and carboxylic acids either along or at the end of the polymer chains. These are all further photolysis products of the primarily formed hydroperoxides.

1.3. Polymer stabilisation

The term 'stabiliser' is used to describe a wide range of technological inhibition processes, although specific terms have been coined to describe particular kinds of inhibition when the precise cause is known. For example, 'flame retardant' refers particularly to the fire situation whereas 'antioxidant' is normally limited to polymers at lower temperatures during service. The terms 'antidegradant' or 'antideteriorant' are frequently used as omnibus descriptions when several mechanisms of stabilisation are involved, or where the mechanism of stabilisation is not understood. Until recently this applied to polymers subjected to the outdoor environment, where the phenomenological term 'weathering' is often used to describe the effects of light in combination with other environmental factors on the degradation of polymers. Ultra-violet (UV) stabilisers for polymers were therefore considered to be mechanistically distinct from antioxidants. However, as the understanding of the mechanism of antioxidants has developed, it has been recognised that UV stabilisers are essentially anti-oxidants.

Rubber technologists use the all-embracing term 'antidegradant' to encompass heat stabilisers, antifatigue agents and antiozonants, but specific antioxidant mechanisms can again be distinguished within these categories.

Antioxidants and stabilisers are essential to the commercial applications of many polymers and the world-wide sales of antioxidants and stabilisers amount to hundreds of thousands of tons per annum. The importance of their use in polypropylene has already been mentioned. A similar but even more striking example of the use of antioxidants is in the motor car tyre. It is true to say that the rubber tyre in its present form could not have been developed to the highly sophisticated engineering product that it is today but for the parallel development of antioxidants to protect it from the effects of the high temperatures and complex mechanical stresses to which it is subjected at high speeds under modern motorway conditions. The development of the radial tyre with its high tread mileage has, somewhat ironically, increased the demand made on the antidegradant 'package' used in the carcass of the tyre. This has also been exacerbated by the trend toward retreading of tyres in the interests of material conservation. Retreading has been practised in truck tyre technology for many years, and the high incidence of failure of large tyres which may operate for considerable periods at temperatures over 100 °C is evidenced by the debris on the hard shoulders of the motorways. One of the factors known to be contributory to this problem is the removal of antioxidants by leaching by road surface water. The practical consequence of this purely physical effect is that the stabiliser package may be substantially depleted at the end of the 'first life' of a tyre, so that a retreaded tyre is virtually unprotected against heat ageing and fatigue. This stimulated research into polymer-bound anti-degradants that are an integral part of the rubber molecule. This topic will be discussed in more detail in section 5.3.3.

Another area of practical importance where loss of antioxidants and UV stabilisers occurs by volatilisation or leaching is in synthetic textile fibres and rubber threads. Due to the high surface area to volume ratio of fibres, antioxidants which can migrate through the polymer are rapidly removed from the surface in contact with dry cleaning solvents or aqueous detergents. Since an important application for polypropylene is in industrial fibres and ropes, the phenomenon of antioxidant loss severely limits the lifetime of such products.

1.4. Recycling of polymers

In the mid 1970s, the industrial societies suddenly became aware that the oil reserves upon which the polymer industry is at present based are limited. This was reinforced by the rapid escalation in the price of crude oil and of oil derived products, and resulted in a call from ecologists and environmentalists to reverse the trend toward the throw-away society. Although some of this criticism was misplaced because it ignored the immense social and economic value of plastics in packaging, this attack did serve to concentrate attention on the possibility of recycling waste plastics, and particularly 'disposable packaging' as an alternative to burning them or burying them as is currently practised.

As has already been indicated, some polymers, notably polymethyl-methacrylate and polystyrene of the packaging polymers, can be readily depolymerised by heating, leading to the recovery of a substantial proportion of the original monomers. However, these polymers form only a minor part of the polymer waste stream. The major generic types, notably polyethylene, polypropylene and poly(vinyl chloride), do not yield mono-mer but low molecular weight fragments which have little value other than as fuel. In addition, PVC presents its own particular problem, both in pyrolysis and on combustion, due to the formation of highly-corrosive HCl gas which rapidly corrodes the recycling plant.

Clearly it would be an advantage if the chemical energy which has been expended in the manufacture of the polymer and which is stored in the polymer structure could be conserved by recycling the polymer itself. There are, however, two serious technical problems involved in the reprocessing of plastics waste. The first, which is outside the scope of this book, is the adverse effects of incompatible impurities (other polymers, paper, metal foil, etc.) on the mechanical behaviour of recycled products. The second is the degradation that occurs during the first life-time and subsequent recycling of plastics. This may lead not only to inferior initial properties, but the long-term performance of recycled products may be inadequate for their new use. One of the most important potential applications for recycled plastics is in agricultural buildings, fence posts and related products, where exposure to the outdoor environment is inevitable. The development of cheap, broad-spectrum antioxidants is, therefore, of some importance to the future of plastics recycling for most applications. Since recycled products are in competition with traditional products, they must not be less cost-effective than the latter.

1.5. Degradable polymers and the plastics litter problem

In spite of the real progress that is being made in the conservation of raw materials reserves by recycling of polymers, plastics litter presents a much less tractable problem since it is normally generated where it cannot be economically collected for recycling. Commercial packaging litter on the sea-shore and agricultural packaging litter in the countryside are very visible examples of the non-biodegradability of the common packaging plastics. Whereas seaweed and even wood and metals rapidly degrade to an unobtrusive form, plastics accumulate in the natural environment where they cause discomfort both to the visitor and to indigenous wildlife.

The problem of plastics litter arises from the nature of packaging materials. If they were not impermeable to water and resistant to micro-organisms they would not be useful in packaging. It has been shown in the case of polyethylene that some biodegradation does occur in plastics that have been exposed to light. This however, does not involve the hydrocarbon polymer itself but its oxidation products. Figure 1.1 shows the effect of fungal digestion of polyethylene which has been exposed to air under ambient conditions for varying times. As has already been seen, photo-oxidation produces a variety of carbonyl oxidation products in polyolefins, and these absorb strongly in the region of 1710–45 cm^{-1} (see Fig. 1.1, curves 2 and 3). After exposure to severe biological attack, these are removed completely (curve 5) leaving a polymer which is chemically as 'pure' as the unoxidised polymer (curve 1).

Photo-oxidation is a faster process at ambient temperatures than uncatalysed thermal oxidation and it can be catalysed by a variety of photo-sensitisers. The possibility of producing commercial plastics with limited but reproducible life-times has excited the interest of polymer scientists for some years and processes have been developed to do this reliably by making use of existing knowledge of polymer degradation and stabilisation chemistry. One of the most demanding uses for plastics with a built in self-destruct mechanism is in protective film (mulching) for fast growing fruits and vegetables, and 'time-controlled' degradable mulch is now used on a

Figure 1.1. Infra-red spectra of HDPE films with different histories: 1 with anti-oxidant after standing at ambient temperature for one year; 2 as 1 without anti-oxidant; 3 as 2 after standing for three years; 4 as 3 after treatment with an aerated medium inoculated with cultivated soil; 5 HDPE powder without antioxidant exposed to aerobic biodegradation for two years before moulding to film with exclusion of air. (Reproduced from a doctoral thesis by Dr A. C. Albertsson, with kind permission.)

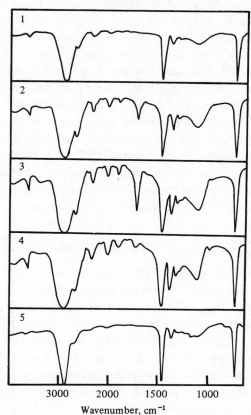

Wavenumber, cm⁻¹

large scale in many countries which rely heavily on automated agriculture for the efficient production of foodstuffs. The chemistry involved in the time-controlled stabilisation of plastics will be discussed in the appropriate sections later in this book.

1.6. Fire hazard of polymers

If there is one factor more than any other that threatens the further replacement of metals and other traditional materials by polymers, it is their flammability. There is evidence that the kinds of regulation which new materials will have to satisfy in the future will be much more stringent than those applied to natural products in the past. The problem is a difficult one for a number of unrelated reasons. In the first place, polymers are organic materials and almost all of them burn if the temperature is high enough. Secondly, the threat to life caused by fire is not only the burning process itself but the toxic effects of degradation products. Thirdly, vapour phase oxidation, which is the essential nature of the burning process, is much more difficult to inhibit than oxidation in the solid polymer. Indeed, a major problem in the use of fire-retarded polymers is that they are frequently so modified by additives that their physical characteristics may be quite different from those of unmodified polymers.

The complex chemistry of flame retardants in the fire situation is at present much less well understood than the chemistry of antioxidants at lower temperatures. The effect of heat on polymers, which forms one aspect of the total combustion process, is now well understood, but the way in which oxygen and flame retardants affect the subsequent chemistry, both in the gaseous phase and in the solid phase, awaits the development of new techniques to identify the reactive chain-carrying species involved.

1.7. The techniques used in the study of polymer degradation and stabilisation

The investigation of polymer degradation has been traditionally carried out at two different levels. The polymer technologist is primarily concerned with the change in physical properties of polymers with time. The polymer scientist, on the other hand, is interested in the reasons for these changes. However, these activities cannot be divorced. Indeed, if further major advances are to be made in the improvement of the durabilities of polymers in service, a better scientific understanding will be necessary of the chemical and physical phenomena involved in polymer degradation and stabilisation under technological conditions.

1.7.1. Technological testing procedures

The ultimate criterion of the durability of a polymer component is the length of time it continues to perform satisfactorily under service conditions. The design of polymeric materials for durability is therefore just as important a design characteristic as dimensional design and the two are frequently interrelated.

In some applications, for example in the use of polymers in short-term uses such as crop protection film (mulching film), the service life may be measured directly. Figure 1.2 shows the change in elongation at break of low density polyethylene film containing a photo-activator with time of out-door exposure in England. Under these conditions, normal polyethylene film would take five or six years to reach the same state of degradation, and UV-stabilised film would take very much longer. It is not normally feasible therefore to await the results of field tests in selecting photostable polymers. A PVC window frame would need to be exposed to the out-door environment for up to 30 years to assess its durability and this would virtually rule out the development of new materials for long-term use. A compromise solution to this problem is the development of 'accelerated ageing' tests. These are intended to accelerate only those

Figure 1.2. Change in elongation at break of LDPE films during out-door exposure in England: 1 without additive; 2 with photo-activator.

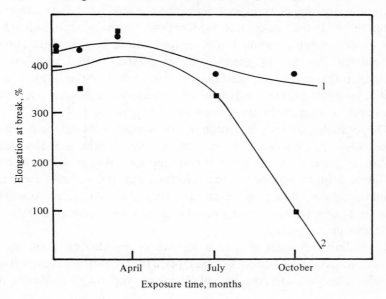

elements of the environment that are destructive to the polymer under service conditions without introducing new ones. The interpretation of accelerated tests and their correlation with service tests thus lies at the centre of the problem of selecting polymers for durability. If accelerated tests are to have relevance to real conditions, this demands an understanding of the physical and chemical phenomena involved in polymer deterioration. For example, it is quite unjustifiable to subject polymers to wavelengths less than 285 nm (the lower limits of the Sun's spectrum on the Earth's surface). These wavelengths certainly accelerate the photo-oxidation of polymers, but they do so unselectively by photolysing chemical bonds which are not broken in sunlight. The consequent chemistry is quite different and irrelevant to the 'weathering' of polymers.

A commonly used accelerated test for oxidative stability is the 'air oven test'. This involves subjecting a suitably fabricated polymer sample (generally either a film or a moulded strip) to temperatures ranging from 70 °C to 150 °C with an air flow passing over the surface of the sample. The change in stress–strain behaviour (tensile strength, elongation at break or modulus at a given extension) is measured on samples removed from the oven at intervals until the specimen is deemed to have 'failed'. The failure point of course depends on the purpose for which the polymer is intended. In the case of the agricultural film exemplified in Fig. 1.2, this is the time at which the film becomes brittle and fragments under slight pressure. Expressed more scientifically, this is the point at which the polymer has lost a certain proportion of its original elongation at break, which is a measure of its toughness. In the case of mulching film this is normally when the film has lost 80–90% of its elongation. For other applications, however, for example polyethylene for use in greenhouses or plastic buckets, the loss in elongation that can be tolerated may be much less than this or even zero. Falling weight impact tests which measure energy to fracture are often used to provide a complementary assessment of toughness.

The technological criteria used in the assessment of the deterioration of some polymers are less concerned with mechanical behaviour than with change in appearance. Many polymers discolour during ageing due to oxidation or functional group elimination along the polymer chain with the formation of polyconjugated unsaturation (e.g. PVC, ABS, etc.). As will be seen in later chapters, these phenomena can be followed readily by spectroscopic techniques.

The validity of accelerated tests depends on whether extrapolation is possible to service conditions. For example, if the air oven test is carried out at various temperatures, the rates of change of properties should obey the

Arrhenius relationship. Where this has been done it has generally been found that it may do so over a relatively small temperature interval but that extrapolation to temperatures more than 20 °C lower than the lowest temperature of test is generally unsafe. One reason for this is that the sample may undergo a morphological change, thus changing the nature of the oxidising medium. A second possibility is that the rate-controlling process may change over the temperature range. An example of this which has been shown to be very important in practice is the physical loss of an antioxidant or stabiliser from the polymer. The rate of this process may be of dominating significance at high temperatures, particularly in a rapid air stream, but this may be unimportant in ambient environments, or in a closed environment at high temperatures. This principle is illustrated in table 1.1, which compares the failure times of polypropylene samples containing different antioxidants (A–I) at temperatures between 100 and 150°C. It is clear that although all antioxidants become less effective with increase in temperature, some (C, E, I), become relatively more effective with increase in temperature, whereas others (A, D), become relatively less effective. Similar effects have been observed on changing the thickness of the samples, and it may be concluded that physical effects, namely the rates of loss of protective agents by volatilisation, dominate their technological performance. This will be discussed in more detail in chapter 5.

It will be evident from what has been said that the design of accelerated tests should always involve an understanding of the chemistry and physics

Table 1.1. *Effect of temperature on the effectiveness of antioxidants A–I in polypropylene (at constant thickness 0.25 cm).*

Days to failure

100 °C	125 °C	140 °C	150 °C
B 167	B 117	*A 22*	*A 9*
C 183	*A 130*	B 34	B 15
A 286	**C 140**	**C 35**	*D 15*
E 300	**E 187**	*D 53*	C 17
G 387	F 226	**E 57**	H 19
H 466	*D 256*	F 64	F 27
I 507	G 259	G 66	G 30
F 600	**I 288**	H 72	**E 31**
D>600	H 297	**I 74**	I 33

Note: Relatively less effective with temperature increase: printed in italic. Relatively more effective with temperature increase: printed in bold.

of the processes occurring in the polymer. Most accelerated ageing tests are therefore a compromise between rapidity and reliability. In general, the nearer to actual conditions an accelerated test can approach, the more reliable it is as a predictor of performance during service.

Some testing procedures do not need to be accelerated. For example, a melt stabiliser for a polymer during processing and fabrication can be evaluated in a simulated manufacturing process. Even here, however, it is important to appreciate that what happens to the polymer during manufacture may profoundly affect its behaviour in service.

1.7.2. The scientific study of polymer degradation processes

(a) Physical methods

The full range of modern physical methods of analysis, most of which have emerged only in the past 30 years, have been brought to bear on the scientific study of polymer degradation processes. The more important of these may be classified into four groups, namely determination of molecular weight, the techniques of thermal analysis, spectroscopy, and chromatography.

Methods of measurement of molecular weight include viscometry, osmometry, light scattering, ultracentrifuge and, historically the most recent, gel permeation chromatography (GPC) which will be mentioned again later. These methods, especially in combination, can give a great deal more information about a polymer sample than the average size of its molecules but it is not appropriate to discuss this here. Figure 1.3 gives an idealised representation of the kind of changes in molecular weight which may occur in a degrading polymer. Thus pathway A represents a situation in which the polymer chains are being randomly cleaved or scissioned so that little volatile material is produced until late in the reaction. Pathway B represents the situation in which the decrease in molecular weight is proportional to the amount of volatile products and suggests that molecules which are volatile at the degradation temperature are being progressively cleaved stepwise from the chain ends. Finally, pathway C represents a situation in which whole polymer molecules are disappearing from the system. This occurs most frequently when the polymer molecule breaks somewhere along its length to form macroradicals which then depropagate or 'unzip' to form monomer in a reaction which is the exact reverse of the propagation process during polymerisation. Of course in any real situation the experimental points will usually lie between these extremes but clearly measurements of molecular weight can often provide a useful preliminary picture of the general way in which a polymer is breaking down.

Thermal analysis is a general description of a family of techniques in which some physical property of a material is continuously recorded while the material is heated, usually through a linear temperature programme and frequently at between 1 and 20° min^{-1}. Those most commonly applied in the study of polymer degradation are thermal gravimetry (TG) in which the loss of weight is recorded, differential scanning calorimetry (DSC) and differential thermal analysis (DTA) in which thermal changes within the polymer due to physical (glass transition or melting temperature) or chemical (degradation) processes are recorded, and thermal volatilisation analysis (TVA) which records the evolution of volatile products. Thermal analysis techniques will be discussed more fully in the following chapter.

Studies in polymer degradation have also made extensive use of the various spectroscopic techniques. Among the most widely applied, the first was probably UV spectroscopy for which commercial instruments began to become available in the late 1940s. This was followed by infrared (IR) spectroscopy in the early 1950s and more recently by nuclear magnetic resonance, mass and electron spin resonance spectroscopy. Certain types of degradation processes in which external chemical substances are involved, such as oxidation, ozonisation and attack by atmospheric pollutants like

Figure 1.3. Idealised representation of changes in molecular weight which may occur in a degrading polymer.

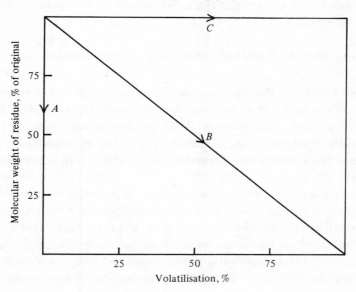

sulphur dioxide or nitrogen dioxide or by other chemicals, are frequently concentrated at the surface of the polymer. Although the total amount of chemical change is small, that in the surface layers may be extensive, causing profound changes in the optical, electrical or mechanical properties of the polymer. The technique of multiple internal reflectance infrared spectroscopy (MIRS) allows a very thin surface layer to be examined to the exclusion of the bulk of the polymer and its development as a standard technique was due in substantial measure to the requirements of the polymer scientist.

Finally, polymer degradation studies have made extensive use of the various forms of chromatography for the separation, identification and quantitative analysis of their products of reaction. These techniques, such as gas liquid chromatography (GLC) and thin layer chromatography (TLC) were developed by organic chemists and later incorporated into the polymer scientist's armoury. Gel permeation chromatography (GPC) was, however, developed primarily as a separation technique in polymer science, making it possible to fractionate a polymer sample according to molecular size much more efficiently and rapidly than hitherto. The basic principle is that small molecules can fit into the pores of a gel much more easily than large molecules so that if a solution of a heterodisperse mixture flows over a suitable gel, the large molecules, being held up least effectively, will be eluted first. Quantification of changes in molecular weight distribution during a degradation process can often be applied to the elucidation of its mechanism. GPC also constitutes a rapid method for the determination of molecular weight.

(b) Oxygen absorption and the formation of oxygen-containing functional groups

Since most polymers degrade in service by oxidation, the kinetics of their reaction with oxygen or of the formation of functional groups provides a valuable and often rapid means of following oxidation resistance. This is a particularly convenient way of studying the effects of antioxidants and stabilisers, since their primary function is to interfere with the oxidation process.

Figure 1.4 shows a typical oxygen absorption curve for a hydrocarbon polymer. It is characterised by an auto-accelerating period which is followed by a rapid quasi-linear rate of oxygen absorption. During the initial stages of an oxidation, molar hydroperoxide concentration as measured by iodimetry follows the same curve as molar oxygen absorption. This follows from the fact that hydroperoxides are the primary products of autoxidation. However, hydroperoxides are unstable and break down at quite moderate temperatures to give carbonyl compounds. The chemistry of this process will be discussed in some detail in later chapters, but it does lead to the third and, in many ways, the most useful chemical method of following the oxidative degradation of polymers, since carbonyl groups can be readily monitored by IR spectrometry (absorbances at 1710–35 cm^{-1}). Some typical IR spectra have already been given in Fig. 1.1, and although this procedure is not very useful from a kinetic standpoint, since the absorbance referred to is a composite one and is difficult to quantify, it does provide useful information on the length of the induction period before oxidation commences in inhibited polymers. It has been found that there is normally a good correlation between the carbonyl index* of hydrocarbon polymers and their physical deterioration, for example change in toughness. This is shown typically for the photo-oxidation of stabilised and unstabilised acrylonitrile–butadiene–styrene copolymer (ABS) in Fig. 1.5. This polymer is particularly sensitive to oxidation due to the presence of the

Figure 1.4. Absorption of oxygen and formation of hydroperoxide (ROOH) and carbonyl compounds (C=O) in high-impact polystyrene during photo-oxidation.

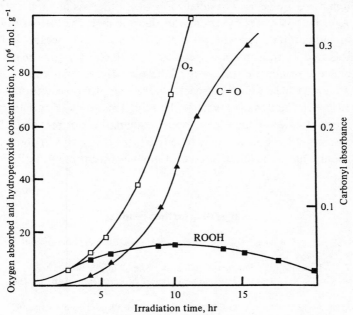

* Carbonyl index is the ratio of the carbonyl absorbance and an invariant absorbance characteristic of the polymer (e.g. C–H stretch). The use of an index compensates for variations in the thickness of the polymer films.

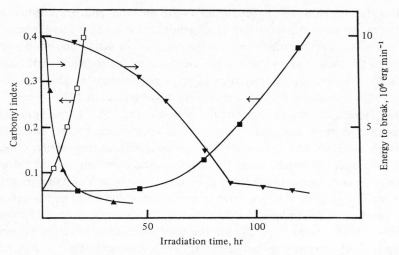

Figure 1.5. Change in impact strength (▲ and ▼) and carbonyl concentration (□ and ■) with UV irradiation time during the photo-oxidation of unstabilised and stabilised acrylonitrile-butadiene-styrene copolymer (ABS).

polybutadiene phase which acts as an oxidation sensitiser in the polyblend. Although there is a general parallelism between the loss of impact strength and the growth of carbonyl groups in the polymer, the former does not show an induction period whereas the latter does. This is because carbonyl was measured by transmission spectroscopy which averages the carbonyl absorbance through the thickness of the film. However it can be shown qualitatively by MIRS infrared spectroscopy that photo-oxidation is much more severe in the surface layers of the film. This leads to surface micro-crack formation at an early stage, thus reducing impact resistance before any measurable oxidation has occurred in the polymer bulk. This phenomenon has to be taken into account when interpreting spectroscopic data.

Suggested further reading

1. G. Scott, *Atmospheric Oxidation and Antioxidants*, Elsevier, London and New York, 1965.
2. W. L. Hawkins (ed.), *Polymer Stabilisation*, John Wiley & Sons, 1972.
3. N. S. Allen (ed.), *Degradation and Stabilisation of Polyolefins*, App. Sci. Pub., London, 1983, chapters 1, 3, 5, 6 and 7.
4. J. Guillet (ed.), *Polymers and Ecological Problems*, Plenum Pub. Corp., 1973.
5. D. Gilead and G. Scott, Time-controlled Stabilisation of Polymers, *Developments in Polymer Stabilisation 5*, ed. G. Scott, App. Sci. Pub., London 1982, chapter 4.
6. G. Scott, Substantive Antioxidants, *Developments in Polymer Stabilisation 4*, ed. G. Scott, App. Sci. Pub., London Chapter 6.

2

Thermal degradation

2.1. Introduction

It is reasonable to anticipate that the thermal stabilities and decomposition reactions of polymers should be similar to those of small molecular compounds with similar chemical constitutions. A low molecular weight alkane like hexadecane(I) for example, should be an excellent model for polyethylene(II), 2,4,6-trichloroheptane(III) for poly(vinyl chloride)(IV) and the ethyl ester of 2-methyl, 2-carboxy butyric acid(V) for poly(ethyl methacrylate)(VI).

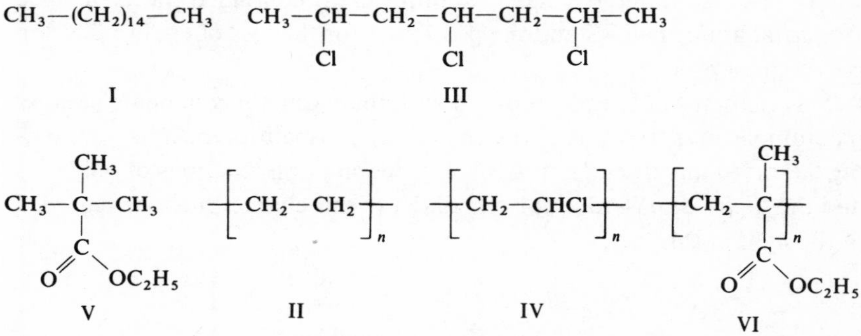

$$CH_3-(CH_2)_{14}-CH_3 \qquad CH_3-\underset{\underset{Cl}{|}}{CH}-CH_2-\underset{\underset{Cl}{|}}{CH}-CH_2-\underset{\underset{Cl}{|}}{CH}-CH_3$$

$$\text{I} \qquad\qquad\qquad\qquad \text{III}$$

$$CH_3-\underset{\underset{\underset{O}{\diagup}\quad\diagdown OC_2H_5}{C}}{\overset{\overset{CH_3}{|}}{\underset{|}{C}}}-CH_3 \qquad \left[CH_2-CH_2\right]_n \qquad \left[CH_2-CHCl\right]_n \qquad \left[CH_2-\underset{\underset{\underset{O}{\diagup}\quad\diagdown OC_2H_5}{C}}{\overset{\overset{CH_3}{|}}{\underset{|}{C}}}\right]_n$$

$$\text{V} \qquad\qquad\qquad \text{II} \qquad\qquad\qquad \text{IV} \qquad\qquad\qquad \text{VI}$$

However, while I and II both yield a mixture of saturated and terminally unsaturated hydrocarbons and III and IV both give HCl as a primary decomposition product, the polymers decompose at very much lower temperatures than the models, perhaps by as much as 200 °C. On the other hand, V and VI give quite different products. The former undergoes ester decomposition yielding ethane and the parent acid while the polymer gives high yields of monomer.

There are two general reasons why the thermal degradation behaviours of polymers and model compounds diverge in these ways. First, although the structures of polymers are normally represented graphically by formulae of the type II, IV and VI, i.e. simply as the monomer unit repeated many times, they are in fact more complex than this. For example, they must have different terminal structures which depend upon their mode of preparation. They may also incorporate chain branches, unsaturated structures and a variety of other features formed as an integral part of the polymerisation process or due to impurities in the polymerising mixture. Structural changes may also have occurred in the polymer during its shelf life, often as a result of the combined effects of sunlight and air. These 'abnormal' structures may then act as centres for the initiation of degradation processes in polymers thereby making them less stable than anticipated.

Secondly, the long-chain character of polymer molecules makes it possible for types of reaction to occur which are impossible in small molecules. Thus monomer production may occur by the 'unzipping' of monomer molecules from a reactive chain end. Alternatively the products of decomposition of one monomer unit may activate the decomposition of an adjacent unit so that the reaction proceeds from unit to unit along the chain. Thus a simple reaction in a small molecule is converted to a rapid chain process in a polymer. The instability of poly(vinyl chloride), which will be discussed later, is a classic example of the combined effects of labile structural abnormalities and the possibility for the loss of HCl to develop into a chain process.

This chapter has been introduced with these general comments because macromolecular structure and structural abnormalities must be dominating factors in any discussion of thermal degradation reactions of polymers and the topics discussed in the remainder of this chapter must be regarded in the light of this fact.

2.2. Experimental methods

In the early literature of polymer degradation, in the period 1945–55, a number of methods were described for following polymer degradation reactions continuously. These usually involved direct measurement of loss in weight or of the evolution of volatile degradation products. Since that time a large number of commercial instruments have evolved which measure these or certain other properties of polymers as they are heated isothermally or under linear temperature programmed conditions. These experimental methods are grouped together under the description 'thermal analysis'. The three most important from the point of view of polymer degradation are thermogravimetry (TG), in which the loss in weight is measured, differential scanning calorimetry (DSC) and the closely related

differential thermal analysis (DTA) in which heat absorption or evolution
due to either physical or chemical changes occurring within the polymer is
measured and thermal volatilisation analysis (TVA) which is a particularly
versatile form of evolved gas analysis (EGA) but for which commercial
instruments are not currently available. In TVA the pressure of volatile
degradation products is measured continuously in a continuously
evacuated system, the pressure being a function, although not a linear
function, of the rate of evolution of volatile material. In TVA, unlike most
other thermal analysis methods, the products of degradation are
immediately available for subsequent analysis as a series of well defined
fractions.

The complementary nature of these three thermal analysis techniques is
illustrated in Fig. 2.1, which refers to the thermal degradation of an
equimolar copolymer of 2-bromoethyl methacrylate (VII) and acrylonitrile
(VIII).

$$CH_2=\underset{\underset{VII}{COOCH_2-CHBr}}{\overset{\overset{CH_3}{|}}{C}} \qquad\qquad CH_2=\underset{\underset{VIII}{CN}}{\overset{|}{CH}}$$

heated from ambient temperature to 500 °C at 10 ° min^{-1}. TG

Figure 2.1. TG, DSC and TVA curves for an equimolar copolymer of 2-bromoethyl
methacrylate and acrylonitrile. TVA trap temperatures: —— 0 °C; – – – – 45 °C; ·····
−75 °C; −·−·− − 100 °C; −··−··− − 196 °C. (Reproduced by kind permission of
Polym. Degrad. and Stab., **4**, 173, 1982.

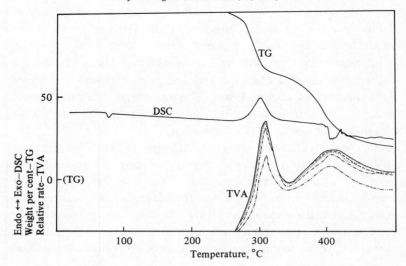

demonstrates that weight loss begins at approximately 250 °C and occurs in two stages. This is reflected in both the DSC and TVA curves. TG also provides information about the amount of weight loss in each step and the amount of residue remaining at 500 °C. On the other hand, DSC indicates that the first stage of degradation is exothermic while the second is endothermic and experience tells us that the irregular shape of the endotherm is probably due to the large volumes of gaseous products bubbling from the viscous liquid polymer. Indirectly this tells us, for example, that cross-linking has probably not occurred up to this point. The various TVA traces provide information about the volatilities of the products of degradation. Thus each curve represents material volatile at the temperatures indicated in the legend. It is clear, therefore, that a very large proportion is volatile at -196 °C while most of the remainder is volatile at -100 °C.

The close correlation which can sometimes be made between TG and TVA data is illustrated by the curves in Fig. 2.2 for a chlorinated rubber containing 64.5% of chlorine. The first peak is due entirely to HCl so that the O°, $-45°$, $-75°$ and -100 °C curves are coincident. Weight loss of 67% indicated by TG suggests that HCl loss is quantitative.

It is important to remember that TG and TVA only provide evidence of reactions which are proceeding with evolution of volatile material. But many serious deteriorative reactions may occur in polymers at very much lower temperatures than those at which volatile products appear. In polystyrene, for example, a decrease in molecular weight occurs at 200 °C although volatile products are not formed up to at least 280 °C. In polyacrylonitrile, on the other hand, the first manifestation of degradation is discolouration which is due to an intramolecular cyclisation process which will be discussed in a later section of this chapter. Evidence of these kinds of reaction may be given by DSC or thermal mechanical analysis (TMA) which monitors a change in physical properties with temperature but for a quantitative chemical study, thermal analysis methods must be supplemented by more quantitative analytical methods like the measurement of molecular weight or IR spectroscopy.

It is also important to remember that although TG and TVA, especially in combination, can give a great deal of information about thermal degradation reactions in polymers, this information is only of a very preliminary nature because it does not provide any direct evidence of the chemical nature of the reactions which are occurring. Thus detailed analysis of products must be made using, as appropriate, one or more of the large number of analytical techniques which have become available to us during the past few decades, especially the various chromatographic and spectroscopic methods.

This detailed analysis of degradation products can usually be assisted by their separation into a series of well-defined fractions which most typically comprise, A products not condensed at -196 °C, B products volatile at ambient temperature but condensed at -196 °C, C products volatile at degradation temperatures but condensed at ambient temperature, and D involatile residue.

Fraction A must comprise one or more simple molecules like hydrogen, carbon monoxide and methane which can be easily identified and quantitatively estimated by, for example, infra-red and mass spectrometry. The upper limit of the molecular weights of the constituents of fraction B will usually be of the order of 150. It is often a mixture of a number of chemically different products which must be separated and analysed usually by the application of appropriate chromatographic and spectroscopic methods. McNeill has recently devised a very simple but effective

Figure 2.2. Simultaneous TG and TVA data for chlorinated natural rubber. TVA trap temperatures: —— 0 °C, -45 °C, -75 °C and -100 °C; $-\cdot-\cdot-$ -196 °C. (Reproduced by kind permission of *Developments in Polymer Degradation – 1*, ed. N. Grassie, App. Sci. Pub., London, 1977, p. 59.)

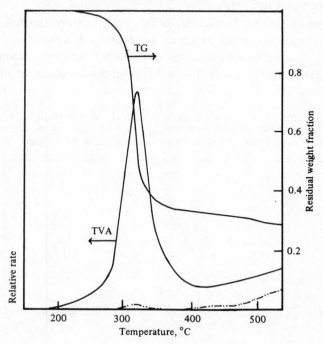

procedure which is often sufficient for the separation and which he describes as sub ambient thermal volatilisation analysis (SATVA). Fraction B is first condensed at − 196 °C under vacuum. The cold bath is transferred to a second trap in the apparatus and the flow of vapour from one trap to the other as the temperature of the first slowly rises is monitored continuously by a pressure gauge. Figure 2.3 shows a SATVA trace for fraction B from the degradation of an equimolar copolymer of 2-bromoethyl methacrylate and styrene. The products have been identified as follows: 1, ethylene; 2, carbon dioxide; 3, hydrogen bromide; 4, vinyl bromide and acetaldehyde (minor); 5, styrene, toluene (trace), benzene (trace); 6, 2-bromoethyl methacrylate.

Fraction C is often described as the 'cold ring' fraction because it appears as a deposit on the cold surface just outside the hot zone of the degradation apparatus. It may be an involatile liquid, a wax or a solid and usually comprises recognizable fragments of the polymer chain in which the monomer units may or may not be modified chemically. It may often be separated chromatographically into its constituents, its molecular weight estimated by vapour phase osmometry, mass spectrometry or gel permeation chromatography (GPC) and IR spectral measurements will usually give an indication of its relationship to the volatile products in fractions A and B.

Depending upon solubility, molecular weight and/or spectral measurements are usually carried out on the residual fraction D.

As much information of this kind as possible must be accumulated if a comprehensive picture of the thermal degradation reactions which occur in

Figure 2.3. SATVA trace for fraction B from the degradation of an equimolar copolymer of 2-bromoethyl methacrylate and styrene. (Reproduced by kind permission of *Polym. Degrad. and Stab.*, **4**, 123, 1982.)

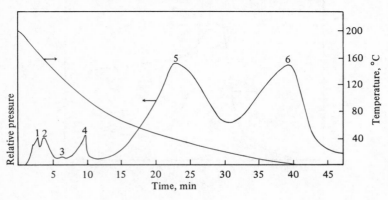

a given polymer system is to be assembled. Although all the detailed evidence will not usually be given, it should be understood that it is by methods of this kind that the information presented in the remainder of this chapter has been obtained.

There are two additional experimental aspects of thermal degradation which must be mentioned briefly because they have been given considerable attention in the literature from time to time. These are flash pyrolysis and the derivation of kinetic parameters from thermal analysis data.

In flash pyrolysis the polymer is raised very rapidly, in a matter of seconds or less, to a relatively high temperature, say 500 °C or more, at which the molecules are broken down into small fragments. The pattern of products is characterised, usually by chromatographic and mass spectrometric methods. Flash pyrolysis is particularly useful for the rapid identification of previously well characterised materials or for the differentiation of polymers of similar structure and it can also give valuable information about degradation mechanism. However it is rather reminiscent of the sledge hammer approach to cracking a nut, which frequently makes it difficult to deduce the precise mechanism of initiation and early stages of breakdown which are important in understanding the deterioration and aging processes in polymers. Just as it is more informative botanically and more valuable commercially to dissect the nut using more sophisticated tools, so is it more informative to dismantle the molecule step by step using milder experimental conditions.

Kinetic parameters, especially orders of reaction and energies of activation, have been deduced by mathematical analysis of thermal analysis data, most frequently temperature programmed TG curves. Like those from flash pyrolysis, results of this kind can be of considerable comparative analytical value, but their interpretation should not be carried too far. Most thermal degradation reactions are complex processes which usually occur in a number of more or less well defined steps. In addition, the physical and chemical nature of the medium is changing as the reaction proceeds and the temperature is raised. Thus kinetic parameters obtained in this way usually bear no direct quantitative relationship to the primary degradation processes which are occurring.

2.3. Classification of thermal degradation reactions

Many different kinds of degradation reactions may be induced thermally in polymers but it is convenient to divide them into two main classes, namely, depolymerisation and substituent reactions. Depolymerisation is characterised by the breaking of the main polymer chain backbone so that at any intermediate stage the products are similar to the parent material in the sense that the monomer units are still distinguishable. The ultimate product may be monomer, as from poly(methyl methacrylate) or volatile chain fragments like the range of short chain alkanes and alkenes from polyethylene.

In substituent reactions it is the substituents attached to the backbone of the polymer molecules which are involved so that the chemical nature of the repeat unit is changed although the chain structure may remain intact. Volatile products, if they are produced, will be quite unlike monomer. Typical examples are the liberation of hydrogen chloride from poly(vinyl chloride)(IV) and cyclisation, reaction (2.1), which occurs in polyacrylonitrile.

$$
\begin{array}{c}
\text{CH}_2 \quad \text{CH}_2 \quad \text{CH}_2 \\
\text{---CH} \quad \text{CH} \quad \text{CH} \quad \text{CH---} \\
| \quad | \quad | \quad | \\
\text{C} \quad \text{C} \quad \text{C} \quad \text{C} \\
||| \quad ||| \quad ||| \quad ||| \\
\text{N} \quad \text{N} \quad \text{N} \quad \text{N}
\end{array}
\longrightarrow
\begin{array}{c}
\text{CH}_2 \quad \text{CH}_2 \quad \text{CH}_2 \\
\text{---CH} \quad \text{CH} \quad \text{CH} \quad \text{CH} \\
\text{C} \quad \text{C} \quad \text{C} \quad \text{C} \\
\text{N} \quad \text{N} \quad \text{N} \quad \text{N}
\end{array}
\qquad (2.1)
$$

2.3.1. Radical depolymerisation reactions

Poly(methyl methacrylate)

The simplest depolymerisation reaction is that of poly(methyl methacrylate)(X) from which monomer is obtained in quantitative yield. Typical TVA curves for both radical- and anion-initiated polymers are shown in Fig. 2.4. In the case of the radical-initiated polymer, the second peak is always the major peak and occurs at the same temperature as the ionic polymer peak. The lower temperature peak in the case of the radical polymer is variable in intensity, depending upon the conditions of polymerisation; e.g. the initiator used and its concentration, the temperature of polymerisation, the presence of transfer agents, etc. The separation of curves in both phases of the reaction is similar, suggesting that the products are closely similar and indeed analysis demonstrates that monomer is the only product.

It has been shown that depolymerisation is a radical chain reaction and that degradation in the region of 300–400 °C is associated with initiation by scission of the polymer at random along its length to form radicals. These radicals then 'depropagate' or unzip in a reaction which is the exact reverse of the propagation process in polymerisation reaction (2.2). Termination may occur by interaction of pairs of radicals as in polymerisation, or by the unzipping process reaching the end of the chain, the small residual radical escaping into the gas phase.

$$\text{~CH}_2-\underset{\underset{\text{COOCH}_3}{|}}{\overset{\overset{\text{CH}_3}{|}}{C}}-\text{CH}_2-\underset{\underset{\text{COOCH}_3}{|}}{\overset{\overset{\text{CH}_3}{|}}{C}}-\text{CH}_2-\underset{\underset{\text{COOCH}_3}{|}}{\overset{\overset{\text{CH}_3}{|}}{C}}\text{~} \longrightarrow$$

X

$$\text{~CH}_2-\underset{\underset{\text{COOCH}_3}{|}}{\overset{\overset{\text{CH}_3}{|}}{C}}-\text{CH}_2-\underset{\underset{\text{COOCH}_3}{|}}{\overset{\overset{\text{CH}_3}{|}}{C}}\cdot \qquad \cdot\text{CH}_2-\underset{\underset{\text{COOCH}_3}{|}}{\overset{\overset{\text{CH}_3}{|}}{C}}\text{~}$$

$$\downarrow$$

$$\text{~CH}_2-\underset{\underset{\text{COOCH}_3}{|}}{\overset{\overset{\text{CH}_3}{|}}{C}}\cdot \quad + \quad \text{CH}_2=\underset{\underset{\text{COOCH}_3}{|}}{\overset{\overset{\text{CH}_3}{|}}{C}} \qquad (2.2)$$

In radical polymerisation at least some termination occurs by disproportionation, reaction (2.3)

$$\text{~CH}_2-\underset{\underset{\text{COOCH}_3}{|}}{\overset{\overset{\text{CH}_3}{|}}{C}}-\text{CH}_2-\underset{\underset{\text{COOCH}_3}{|}}{\overset{\overset{\text{CH}_3}{|}}{C}}\cdot \quad + \quad \cdot\underset{\underset{\text{COOCH}_3}{|}}{\overset{\overset{\text{CH}_3}{|}}{C}}-\text{CH}_2\text{~} \longrightarrow$$

$$\text{~CH}_2-\underset{\underset{\text{COOCH}_3}{|}}{\overset{\overset{\text{CH}_3}{|}}{C}}\overset{!}{+}\text{CH}_2-\overset{\overset{\text{CH}_2}{\|}}{C} \quad + \quad \text{HC}-\underset{\underset{\text{COOCH}_3}{|}}{\overset{\overset{\text{CH}_3}{|}}{C}}\text{~}$$

$$(2.3)$$

Thus a proportion of molecules have unsaturated chain terminal structures. It follows that the bond indicated is weakened by about 80 kJ which is the resonance stabilisation energy of the allyl radical which would be formed by its scission. Initiation by this pathway then allows a limited amount of degradation at lower temperatures in radical-initiated polymers. This effect of unsaturation in weakening adjacent bonds and thus giving rise to weak links in the chains is probably the most common reason for instability in polymers.

Polystyrene

Thermal analysis shows that polystyrene degrades thermally in a single step and that monomeric styrene (approximately 40%) is the principal volatile product together with very much smaller amounts of benzene and toluene. There is, however, a large cold ring fraction which consists of decreasing amounts of dimer, trimer, tetramer and pentamer. These oligomers are formed in intramolecular transfer reactions, shown typically in reaction

$$\text{\textasciitilde\textasciitilde}CH_2-\underset{\phi}{CH}-CH_2-\underset{\phi}{CH}-CH_2-\underset{\phi}{\dot{C}H} \longrightarrow \text{\textasciitilde\textasciitilde}CH_2-\underset{\phi}{CH}-CH_2-\underset{\phi}{\dot{C}}-CH_2-\underset{\phi}{CH_2}$$

$$\downarrow$$

$$\text{\textasciitilde}CH_2-\underset{\phi}{\dot{C}H} \; + \; CH_2=\underset{\phi}{C}-CH_2-\underset{\phi}{CH_2}$$

$$(2.4)$$

Figure 2.4. TVA curves (0°C curves only) for poly(methyl methacrylate). (*a*) Prepared by a free radical route: —— $M_n = 20\,000$; – – – $M_n = 100\,000$; ····· $M_n = 480\,000$. (*b*) Prepared by an anionic route: – – – $M_n = 60\,000$; ····· $M_n = 480\,000$; —— $M_n = 20\,000$ free radical polymer for comparison. (Reproduced by kind permission of *Developments in Polymer Degradation – 1*, ed. N. Grassie, App. Sci. Pub., London, 1977, p. 47.)

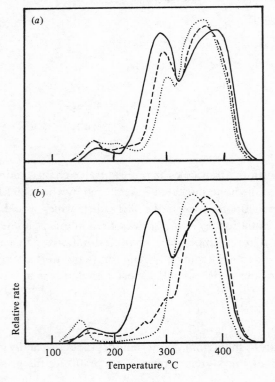

(2.4) which are in direct competition with the monomer-producing de-polymerisation process. By analogy with the description of de-polymerisation as unzipping, these transfer reactions have been described as 'unbuttoning'.

Comparison of the monomer yields from polymers of styrene, α and β-deuterostyrene and α-methyl styrene, presented in table 2.1, clearly show that it is the α hydrogen atom which is involved in the transfer process. Its replacement by deuterium has a strong suppressant effect on transfer while its replacement by a methyl group completely eliminates transfer. Deuteration at the β position or methylation in the ring has no significant corresponding effects. The depolymerisation of polystyrene is thus similar to that of poly(methyl methacrylate) in the sense that as products are formed the same radical is reproduced.

A significant feature of the thermal degradation of polystyrene is the rapid initial decrease in molecular weight, which is followed by a more gradual fall as shown in Fig. 2.5. The reason for the initial rapid decrease has been a matter of controversy for at least 30 years and although it has been studied intensively the final verdict remains to be given. There are two opposing theories. In the first it is suggested that chain scission follows from intermolecular transfer, reaction (2.5), the chain fragments being too large to evaporate from the system even at degradation temperatures.

Table 2.1. *Monomer yields from polymers.*

Polymer	Monomer, %
Methyl methacrylate	100
styrene	42
α-deuterostyrene	70
β-deuterostyrene	42
α-methylstyrene	100
m-methylstyrene	52
ethylene	<1
methacrylonitrile	100
vinylidene cyanide	100
isobutene	32
propylene	2
methyl acrylate	trace
butadiene	1.5
isoprene	12
tetrafluoroethylene	100
chlorotrifluoroethylene	28

$$\text{~CH}_2\text{--}\overset{\displaystyle|}{\underset{\displaystyle\phi}{\text{CH}}} + \text{~CH}_2\text{--}\overset{\displaystyle|}{\underset{\displaystyle\phi}{\text{CH}}}\text{--CH}_2\text{--}\overset{\displaystyle|}{\underset{\displaystyle\phi}{\text{CH}}}\text{--CH}_2\text{--}\overset{\displaystyle|}{\underset{\displaystyle\phi}{\text{CH}}}\text{~} \longrightarrow \text{~CH}_2\text{--}\overset{\displaystyle|}{\underset{\displaystyle\phi}{\text{CH}}}_2 +$$

$$\text{~CH}_2\text{--}\overset{\displaystyle|}{\underset{\displaystyle\phi}{\text{CH}}}\cdot + \text{CH}_2\text{=}\overset{\displaystyle|}{\underset{\displaystyle\phi}{\text{C}}}\text{--CH}_2\text{--}\overset{\displaystyle|}{\underset{\displaystyle\phi}{\text{CH}}}\text{~} \longleftarrow \text{~CH}_2\text{--}\overset{\displaystyle|}{\underset{\displaystyle\phi}{\text{CH}}}\text{--CH}_2\text{--}\overset{\displaystyle|}{\underset{\displaystyle\phi}{\dot{\text{C}}}}\text{--CH}_2\text{--}\overset{\displaystyle|}{\underset{\displaystyle\phi}{\text{CH}}}\text{~}$$

$$(2.5)$$

Alternatively it has been suggested that a limited proportion of 'weak links' are present in polystyrene which break down early in the reaction. A great deal of evidence has been presented for each theory but there now seems no doubt that while a proportion of intermolecular transfer undoubtedly contributes to the decrease in molecular weight later in the reaction, the initial decrease in molecular weight is predominantly due to labile structures which have become incorporated into the polymer molecules during their formation.

It is possible to accentuate chain scission by working at low temperatures at which the evolution of volatile products is very slow. If chain scission

Figure 2.5. Effect of degradation at various temperatures on the molecular weight of polystyrene. ● 280 °C; ■ 290 °C; ▲ 298 °C. (Reproduced by kind permission of *Polymer Science*, ed. A. D. Jenkins, North Holland Publishing Company, 1972, p. 1451.)

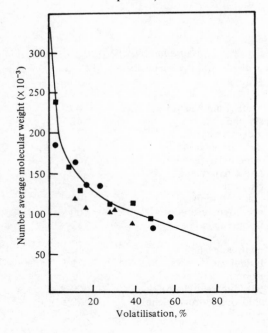

occurs in a polymer molecule in absence of volatilisation, then,

$$P_t = P_0/(s+1) \qquad (i)$$

in which P_0 and P_t are the chain lengths of the polymer initially and after time t at which s scissions have occurred on average per molecule. Thus

$$s = (P_0/P_t) - 1 \qquad (ii)$$

and the fraction of bonds broken, α, is given by equation (*iii*)

$$\alpha = s/P_0 = 1/P_t - 1/P_0. \qquad (iii)$$

If chain scission is random, that is, every interunit bond in every molecule is equally liable to break, then

$$\alpha = kt \qquad (iv)$$

in which k is the rate constant for chain scission. Thus for purely random scission a plot of α against t should be linear and pass through the origin. On the other hand, if molecules contain some weak links which break more rapidly at the beginning of the reaction, then,

$$\alpha = \beta + kt \qquad (v)$$

in which β is the fraction of weak links in the molecules. Figure 2.6 shows that radical polymer obeys equation (*v*) and does indeed incorporate weak links.

The precise chemical structure of the weak links has been more difficult to establish with certainty. Head-to-head linkages and unsaturated structures have been considered and investigated but the current balance of opinion seems to be in favour of oxygenated structures, possibly peroxide linkages formed during the preparation of the polymer by reaction with traces of oxygen.

Polyethylene

The volatile products of degradation of polyethylene are even more complex than those of polystyrene, consisting of an apparently continuous spectrum of hydrocarbons with from 1 to 70 carbon atoms. This is accompanied by a very rapid decrease in molecular weight. Depropagation is no significant since little monomer is produced but a large proportion of mono-olefins is clearly formed in intramolecular transfer processes. A proportion of larger fragments, on the other hand must be formed in intermolecular transfer reactions. Of all the volatile products, propene and 1-hexene are the most abundant and this is undoubtedly due to the fact that reaction of a radical with a hydrogen atom on the fifth carbon atom should be geometrically very favourable since the transition state is a six membered ring.

$$\begin{array}{c} \text{CH}_2\text{---CH}_2 \\ \diagup \qquad \diagdown \\ \text{\textasciitilde CH}_2\text{---CH} \qquad \text{CH}_2 \\ \diagdown \qquad \diagup \\ \text{H} \qquad \text{·CH}_2 \end{array} \longrightarrow \quad \text{\textasciitilde CH}_2\text{---}\overset{}{\text{CH}}\text{---CH}_2\text{---CH}_2\text{---CH}_2\text{---CH}_3 \quad .$$

$$\text{\textasciitilde CH}_2\text{---CH}=\text{CH}_2 + \text{·CH}_2\text{---CH}_2\text{---CH}_3$$

$$\text{\textasciitilde·} + \text{CH}_2=\text{CH---CH}_2\text{---CH}_2\text{---CH}_2\text{---CH}_3$$

$$(2.6)$$

It was shown as early as 1949 that the molecular weight of polyethylene decreases to a limited extent at temperatures much lower than those at which volatile products appear. Four possible weak link structures are known to exist in polyethylene, namely peroxides, carbonyl groups, chain

Figure 2.6. Time dependence of chain scission for a radical initiated polystyrene heated at: ● 300 °C; ▲ 294 °C; ■ 287 °C; ▼ 280 °C. (Reproduced by kind permission of *Polymer Science*, ed. A. D. Jenkins, North Holland Publishing Company, 1972, p. 1465.)

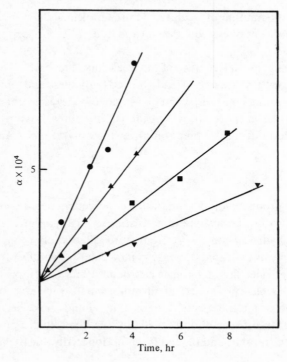

branches and unsaturated structures. One double bond appears for each chain scission but this unsaturation is of three types, namely,

$$R–CH=CH–R', \quad RR'C=CH_2, \quad R–CH=CH_2.$$

The concentrations of the first two soon reach a maximum but the third continues to be formed throughout the reaction. From this and other evidence it may be concluded that the true weak links in polyethylene are probably peroxide groups due to a small amount of oxidation during preparation, storage and processing of the polymer. These then break down to form radicals which abstract hydrogen atoms in unbranched parts of the molecules or at or near chain branches.

$$R–\overset{\cdot}{C}H–CH_2–R' \longrightarrow R–CH\!=\!CH_2 + R'\cdot \qquad (2.7)$$

$$\underset{R'}{\overset{R}{>}}\!\!C–CH_2–R'' \longrightarrow \underset{R'}{\overset{R}{>}}\!\!C\!=\!CH_2 + R''\cdot \qquad (2.8)$$

$$\underset{R'}{\overset{R}{>}}\!\!CH–\overset{\cdot}{C}H–CH_2–R'' \longrightarrow R–CH\!=\!CH–CH_2–R'' + R'\cdot \qquad (2.9)$$

The limited concentration of the more reactive branch points is clearly associated with the rapid initial rate of chain scission and the rapid rise to maximum concentration of the two latter structures.

General mechanism of radical depolymerisation

In spite of the fact that the degradation behaviour of poly(methyl methacrylate), polystyrene and polyethylene are apparently so different, they may be described in terms of a single depolymerisation reaction mechanism,

Random initiation	$M_n \rightarrow M_j^{\cdot} + M_{n-j}^{\cdot}$	(2.10)
Terminal initiation	$M_n \rightarrow M_{n-1}^{\cdot} + M^{\cdot}$	(2.11)
Depropagation	$M_i \rightarrow M_{i-1}^{\cdot} + M$	(2.12)
Intramolecular transfer and scission	$M_i \rightarrow M_{i-z}^{\cdot} + M_z$	(2.13)
Intermolecular transfer	$M_i^{\cdot} + M_n \rightarrow M_i + M_n^{\cdot}$	(2.14)
Scission	$M_n^{\cdot} \rightarrow M_j + M_{n-j}^{\cdot}$	(2.15)
Termination	$M_i^{\cdot} + M_j^{\cdot} \rightarrow M_i + M_j \text{ or } M_{i+j}$	(2.16)

in which n is the chain length of the starting material and M_i, M_j, etc. and M_i^*, M_j^*, etc. represent respectively 'dead' polymer molecules and long chain radicals, i, j, etc. monomer units in length. Information about the initiation and termination processes is available for only a very few systems, but the relative amounts of monomer and non-monomeric products gives some idea of the relative probabilities of depropagation and transfer and allows qualitative correlations to be made between the polymer structure and the nature of the depolymerisation reaction which occurs.

The data in Table 2.1 have already demonstrated the importance of reactive tertiary hydrogen atoms for the production of oligomers while methylene hydrogen atoms play no significant part. Neither do substituents on the benzene rings have much effect. Transfer is completely suppressed when the reactive hydrogen atom is substituted by a methyl group as in poly(α-methyl styrene) and monomer production is typical of polymers of 1,1-disubstituted monomers. Thus monomer is the only significant volatile product from a number (but not all, as we shall see later) of polymethacrylates (IX), polymethacrylonitrile (X) and of poly(vinylidene cyanide) (XI).

$$-CH_2-\underset{\underset{\text{COOR}}{|}}{\overset{\overset{\text{CH}_3}{|}}{C}}- \qquad -CH_2-\underset{\underset{\text{CN}}{|}}{\overset{\overset{\text{CH}_3}{|}}{C}}- \qquad -CH_2-\underset{\underset{\text{CN}}{|}}{\overset{\overset{\text{CN}}{|}}{C}}-$$

$$\text{IX} \qquad\qquad\qquad \text{X} \qquad\qquad\qquad \text{XI}$$

In each of these the degrading radical is relatively unreactive by virtue of being trisubstituted and also in the alpha position with respect to an unsaturated group. Abstraction of primary and secondary hydrogen atoms, which are the only types available, thus cannot compete effectively with depropagation. The stabilising effect of the two methyl groups in polyisobutene (XII)

$$\sim\!\!\sim\!\!CH_2-\underset{\underset{\text{CH}_3}{|}}{\overset{\overset{\text{CH}_3}{|}}{C}}\!\!\sim\!\!\sim \qquad \sim\!\!\sim\!\!CH_2-\underset{}{\overset{\overset{\text{CH}_3}{|}}{CH}}\!\!\sim\!\!\sim \qquad \sim\!\!\sim\!\!CH_2-\underset{\underset{\text{COOCH}_3}{|}}{\overset{}{CH}}\!\!\sim\!\!\sim$$

$$\text{XII} \qquad\qquad\qquad \text{XIII} \qquad\qquad\qquad \text{XIV}$$

is very much less than that of the substituents in IX to XI and the resulting more reactive radical is capable of transfer with primary and secondary hydrogen atoms, with the result that depropagation and transfer products are produced in comparable amounts. The availability of reactive hydrogen atoms is greater in polypropylene (XIII) and radical reactivity increases in the series polyisobutene, polypropylene, polyethylene. The monomer yield becomes negligible as transfer takes over.

The relative radical stabilising effects of the carboxyl group and of the benzene ring are evident by comparing poly(methyl acrylate) (XIV) and polystyrene. In the former the predominance of transfer is complete. In the latter it is only just able to compete with depropagation.

The influence of radical stability is further emphasised by a comparison of the behaviour of polyethylene and polypropylene with polybutadiene (XV) and polyisoprene (XVI)

$$\sim\!\!CH_2\!-\!CH\!=\!CH\!-\!CH_2\!\sim \qquad \sim\!\!CH_2\!-\!\overset{\overset{\displaystyle CH_3}{|}}{C}\!=\!CH\!-\!CH_2\!\sim \qquad \sim\!\!\overset{\overset{\displaystyle F}{|}}{\underset{\underset{\displaystyle F}{|}}{C}}\!-\!\overset{\overset{\displaystyle F}{|}}{\underset{\underset{\displaystyle F}{|}}{C}}\!\sim$$

XV XVI XVII

$$\sim\!\!\overset{\overset{\displaystyle F}{|}}{\underset{\underset{\displaystyle F}{|}}{C}}\!-\!\overset{\overset{\displaystyle F}{|}}{\underset{\underset{\displaystyle Cl}{|}}{C}}\!\sim$$

XVIII

The hydrogen atom availability in polyethylene and polybutadiene is comparable, as it is in polypropylene and polyisoprene, but the predominance of transfer in the saturated as compared to the unsaturated polymers leads to a larger proportion of depropagation as a result of the increased radical stability conferred by the α-unsaturation. The allylic stabilisation is seen to be greater in polyisoprene than in polybutadiene, a measure of the additional stabilising effect of the methyl substituent.

The much greater strength of the carbon–fluorine bond over the carbon–hydrogen bond is reflected in the fact that the transfer process which controls the degradation of polyethylene is virtually absent in the thermal decomposition of polytetrafluoroethylene (XVII) while if a fluorine atom is replaced by chlorine as in polychlorotrifluoroethylene (XVIII) the weaker C–Cl bond allows transfer and the monomer yield drops to 28 %. When hydrogen atoms are introduced into fluorinated monomers, alternative reactions to depolymerisation occur.

2.3.2. *Non-radical depolymerisation reactions*

Many depolymerisation reactions do not involve radicals and three polymer types have been chosen for more detailed consideration because they illustrate some general principles as well as being important as commercial materials. These are polyesters, polysiloxanes and polyurethanes.

Polyesters

Since long chain polyesters can be prepared by the reaction of any dicarboxylic acid with any glycol and since there are large numbers of each of these classes of compounds, it is clear that polyesters with a very large range of structures may be prepared. Both aliphatic and aromatic polymers are widely applied commercially, the former especially in the surface coatings industry and the latter as film products and textile fibres. Although each polyester will have its own detailed mechanism and volatile products of degradation there is a good deal of common ground between them. For example, the major initial step in all cases where β hydrogen atoms are available is invariably alkyl-oxygen scission involving a six-membered ring transition state which, in poly(ethylene terephthalate), is reflected in the formation of

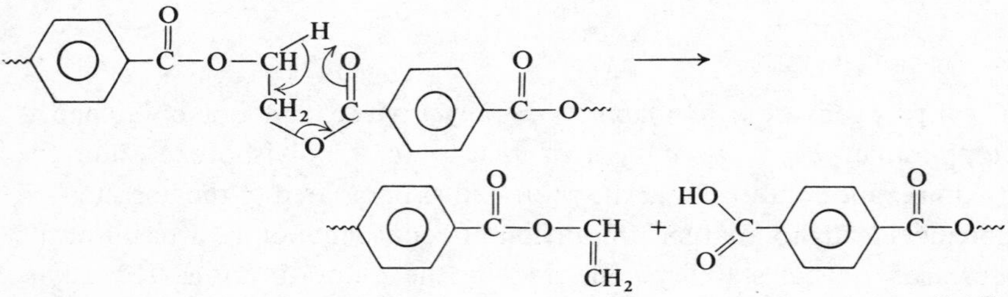

$$(2.17)$$

vinyl and carboxyl end groups. Differences in the ultimate pattern of degradation products are due to differences in the structures of these end groups.

In the aliphatic polyesters the acidic products lead to carbon dioxide, cyclic ketones, cyclic aldehydes and water while vinyl groups tend to give aldehydes, dienes and cyclic ethers. The present discussion will be confined to the aromatic polyester, poly(ethylene terephthalate).

The above ester decomposition reaction occurs at random along the polymer chains and the molecular weight decreases accordingly. Terephthalic acid, acetaldehyde and carbon monoxide are the principal volatile products but the overall reaction is complex and anhydride groups, benzoic acid, *p*-acetyl benzoic acid, acetophenone, vinyl benzoate, ketones, water, methane, ethylene and acetylene have all been detected. The formation of terephthalic acid is easily explained by decomposition at an ester group adjacent to the original scission. On the other hand it has been shown that the vinyl benzoate end group may react in a number of ways; reaction (2.18)

$$\sim C_6H_4 - \overset{\overset{\displaystyle O}{\|}}{C} - O - CH = CH_2 \quad \begin{array}{l} \xrightarrow{\text{Minor}} \quad \sim C_6H_4 - COOH + CH \equiv CH \\[2mm] \xrightarrow{\hspace{1cm}} \quad \sim C_6H_4 - CH = CH_2 + CO_2 \\[2mm] \xrightarrow{\text{Major}} \quad \sim C_6H_4 - \overset{\overset{\displaystyle O}{\|}}{C} - CH_2 - CHO \longrightarrow \end{array}$$

$$\sim C_6H_4 - CO - CH_3 + CO$$

(2.18)

Breaking of appropriate bonds in these structures followed by abstraction of hydrogen atoms will obviously lead to many of the products quoted.

Polysiloxanes

Figure 2.7 shows a typical GLC trace of the volatile products of degradation of poly(dimethyl siloxane) (XIX) which is the basic commercial polysiloxane or silicone.

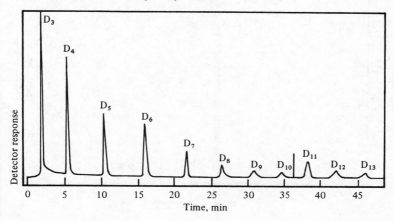

XIX   XX   XXI

Figure 2.7. GLC trace of products of degradation of a poly(dimethyl siloxane). D_3, D_4, etc. refer to cyclic trimer, tetramer, etc. (Reproduced by kind permission of *Europ. Polym. J.*, **14**, 875, 1978.)

These products have been identified as the whole range of cyclic oligomers from trimer (XX) to at least heptadecamer.

The TG curves 1–5 in Fig. 2.8 for a number of polymers give an impression of the stability of pure materials while curve 6 which refers to polymer 2 with 5% added KOH, demonstrates how this compound, which is often used industrially as a polymerisation catalyst, will cause drastic instability. A small amount of methane is also formed in presence but not in absence of KOH.

As normally prepared, these polysiloxanes are terminated by hydroxyl groups but they may be 'end-capped', for example by replacing the hydroxyl groups by trimethylsilyl structures by treating the polymer with hexamethyldisilazane, reaction (2.19).

$$\begin{array}{c} CH_3 \\ | \\ \sim\!\!Si\!-\!OH \\ | \\ CH_3 \end{array} + (CH_3)_3\ Si\!-\!N\!\!=\!\!N\!-\!Si(CH_3)_3 \longrightarrow$$

$$\begin{array}{cc} CH_3 & CH_3 \\ | & | \\ \sim\!\!Si\!-\!O\!-\!Si\!-\!CH_3 \\ | & | \\ CH_3 & CH_3 \end{array} + (CH_3)_3 SiH + N_2$$

$$(2.19)$$

Comparison of curves 3 and 5 in Fig. 2.8 shows how end-capping enhances stability. Curves 1 and 4 also demonstrate how stability increases with

Figure 2.8. TG curves for poly(dimethyl siloxanes). Number average molecular weights: 1, 94 500; 2, 111 500; 3, 183 000; 4, 258 000; 5, 2 end-capped; 6, 2 + 5% KOH. (Reproduced by kind permission of *Europ. Polym. J.*, **14**, 875, 1978.)

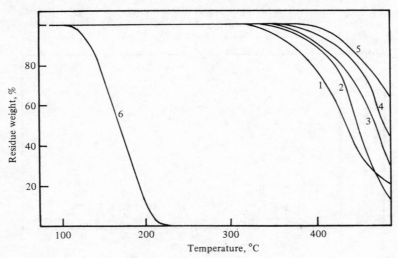

increasing molecular weight. Thus it is clear that terminal hydroxyl groups play some part in the degradation reaction. The following mechanism (reaction (2.20)) is now generally accepted,

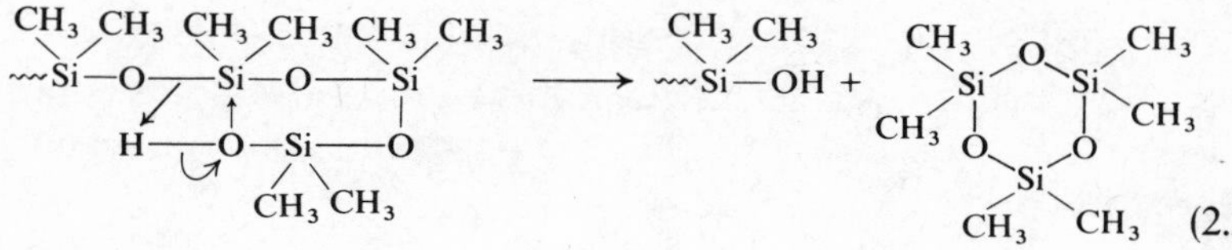

(2.20)

reaction of the hydroxyl groups at points farther from the chain ends leading to higher oligomers.

Although end-capping imparts stability to the polymer the products are the same as for hydroxyl terminated polymers and a comparable mechanism (reaction 2.21) may be suggested in which a second silicon atom

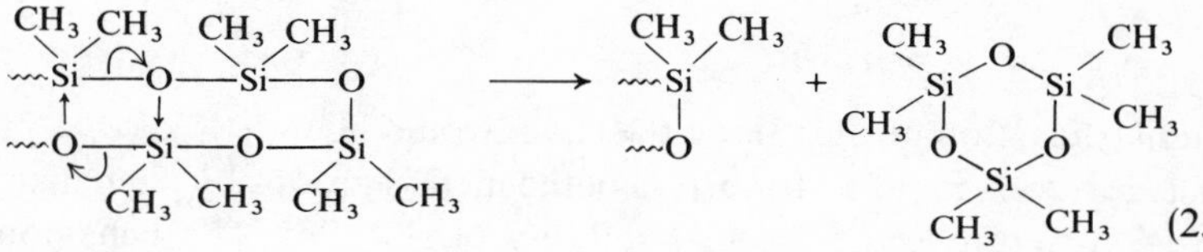

(2.21)

replaces the hydrogen atom in the four-membered ring transition state. Acceleration of the degradation reaction by KOH is probably due to hydroxyl ions providing the driving force for the oligomer-forming reaction.

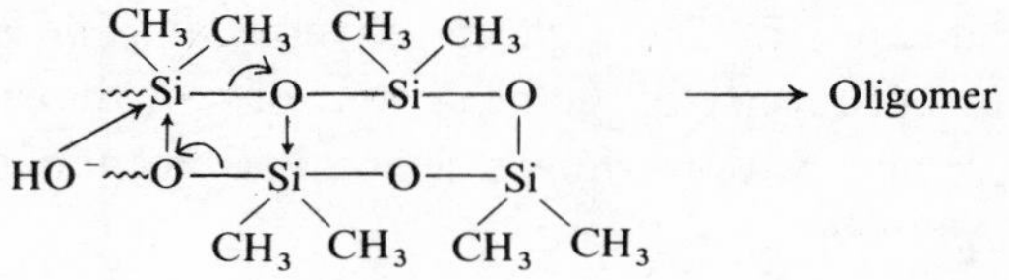

(2.22)

When the reaction is carried out in presence of KOH, mass spectrometry reveals a minor product with a molecular weight of 430. The only reasonable structure corresponds to XXII

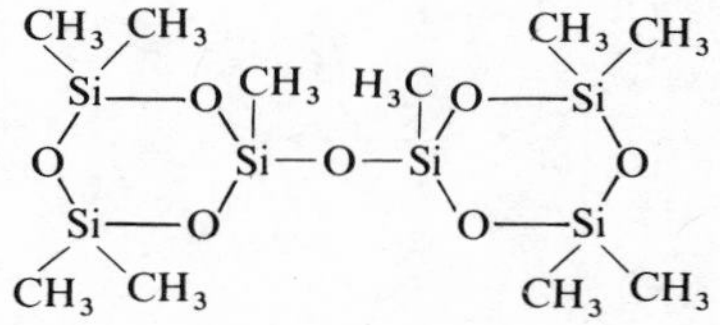

XXII

It has been proposed that the first step in its formation involves a siloxy ion which reacts to form methane, reaction (2.23), which is also formed in the hydroxyl ion catalysed reaction.

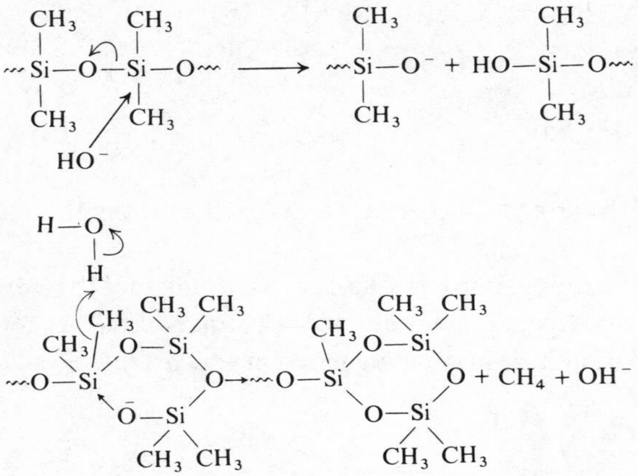

(2.23)

A similar reaction further along the chain would liberate XXII.

The replacement of a proportion of the methyl groups by phenyl groups is used to improve the thermal stability of silicones. The behaviour of poly(methyl phenyl siloxane) (XXI) provides a clue to the reasons for this. Thus benzene is formed in small amounts and unlike poly(dimethylsiloxane) the polymer becomes insoluble on heating even at temperatures below 150 °C. In addition, TG demonstrates that the higher the molecular weight, the lower is the stability and that the stability is decreased by end-capping. These effects of molecular weight and end-capping are the exact reverse of the behaviour of poly(dimethylsiloxane). Nevertheless these observations show a clear involvement of chain ends in the degradation process.

Since hydroxyl terminal structures confer greater stability than trimethylsiloxy structures and since the production of benzene depends upon the number of hydroxyl groups, it has been proposed that the principal effect of hydroxyl groups is to assist the cleavage of Si–Ph bonds, reaction (2.24).

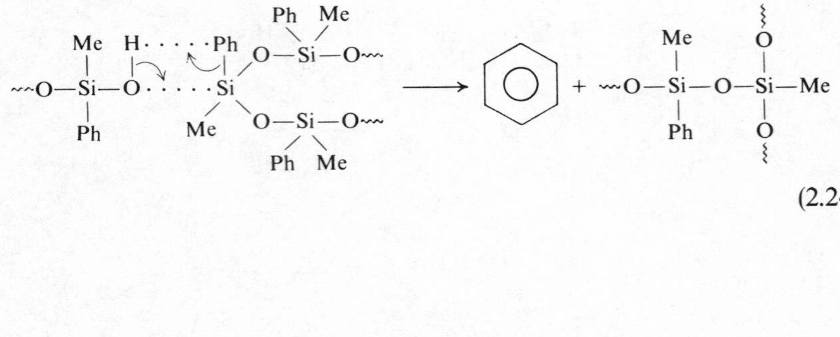

(2.24)

This reaction, which introduces branching into the polymer, will also account for the development of insolubility.

Since the formation of cyclic oligomers is not associated with chain terminal structures, the mechanism proposed for poly(dimethylsiloxane), reaction (2.25), also seems reasonable here.

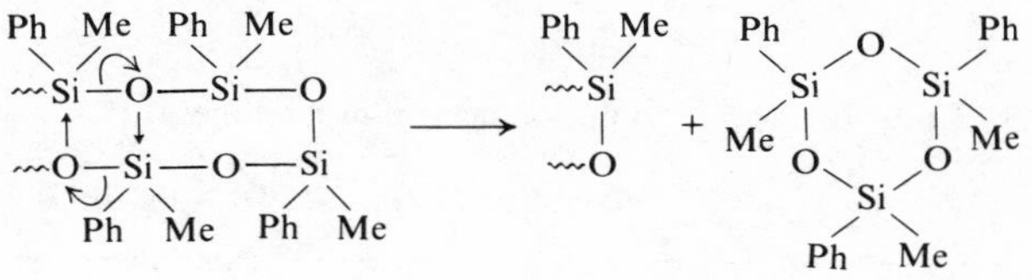

$$(2.25)$$

Polyurethanes

Commercial polyurethanes are usually prepared by reaction of a diisocyanate with a dihydroxy compound which is usually a polyester or a polyether. Products and mechanisms of thermal degradation of these polymers are of particular interest from the point of view of flammability and fire retardance and will be discussed in some detail in chapter 6 from that point of view. At this point it is of interest to focus attention upon the degradation of the polyurethane link and the polymer prepared from 1,4-butane diol and methylene bis-(4-phenyl diisocyanate) has been used for this purpose.

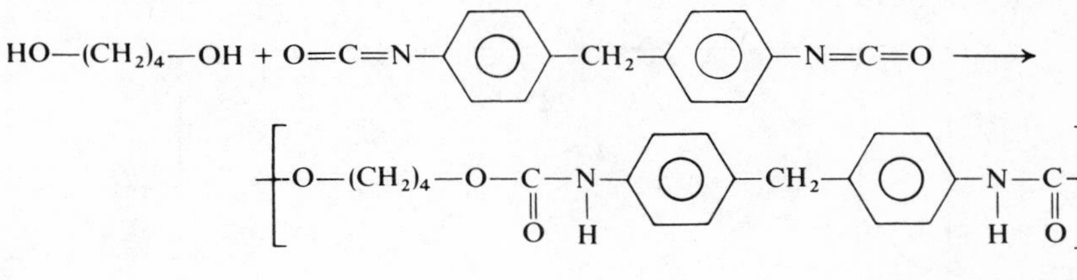

$$(2.26)$$

A TVA trace for this polyurethane is shown in Fig. 2.9. The broad low temperature peak is due to evolution of solvent and the high temperature peak, which shows a maximum rate about $300\,^{\circ}C$ can be attributed to genuine degradation products. A large cold ring fraction is also formed which is in two parts – a liquid and solid whose IR spectra prove that they are the two monomers. A SATVA trace of the more volatile products is shown in Fig. 2.10. The materials in each peak were identified by a combination of IR spectrometry, mass spectrometry and GLC as 1, CO_2; 2, butadiene; 3, tetrahydrofuran; 4, dihydrofuran and 5, water. Carbon monoxide is also formed and a trace of HCN if large samples are degraded.

The IR spectrum of the original polymer is compared in Fig. 2.11 with those of the residues obtained after heating to 300°, 350° and 450 °C at 10° min⁻¹. Profound changes clearly occur but attention is drawn in particular to the following. The disappearance of bands in the vicinity of 3400 cm⁻¹ and 1700 cm⁻¹ which may be assigned to the N–H and

Figure 2.9. TVA trace for a polyurethane. Trap temperatures: —— 0 °C; - - - −45 °C; ····· −75 °C; −·−·− −100 °C; −··−··− −196 °C. (Reproduced by kind permission of *J. Polym. Sci. Chem. Ed.*, **16**, 1563, 1978.)

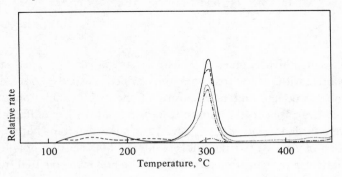

Figure 2.10. SATVA trace for volatile products of degradation of a polyurethane. (Reproduced by kind permission of *J. Polym. Sci. Chem. Ed.*, **16**, 1563, 1978.)

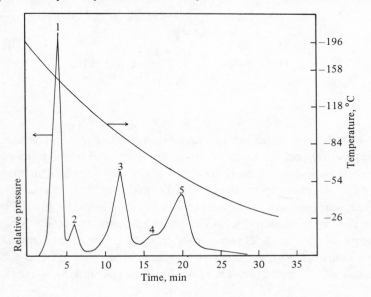

carbonyl groups in the polyurethane, and the appearance of new bands at 2120–2100 cm^{-1} and 1640 cm^{-1} which may be assigned to carbodiimide and urea amide respectively.

The variety of products formed immediately suggests that the mechanism of thermal degradation of polyurethane is very complex. In fact it can be shown spectroscopically that at about $210\,^\circ$C the polyurethane linkage disappears without any of the other products being formed. Thus the overall reaction can be explained in terms of a simple depolymerisation reaction which is the exact reverse of the polymerisation process in which the polyurethane is formed. Thus the two monomers are primarily formed and all the other products are formed from these monomers while they are diffusing from the hot polymer. The overall reaction is represented by the mechanism shown in scheme 2.1 which accounts for all the observed products.

Scheme 2.1 Mechanism of thermal degradation of a polyurethane.

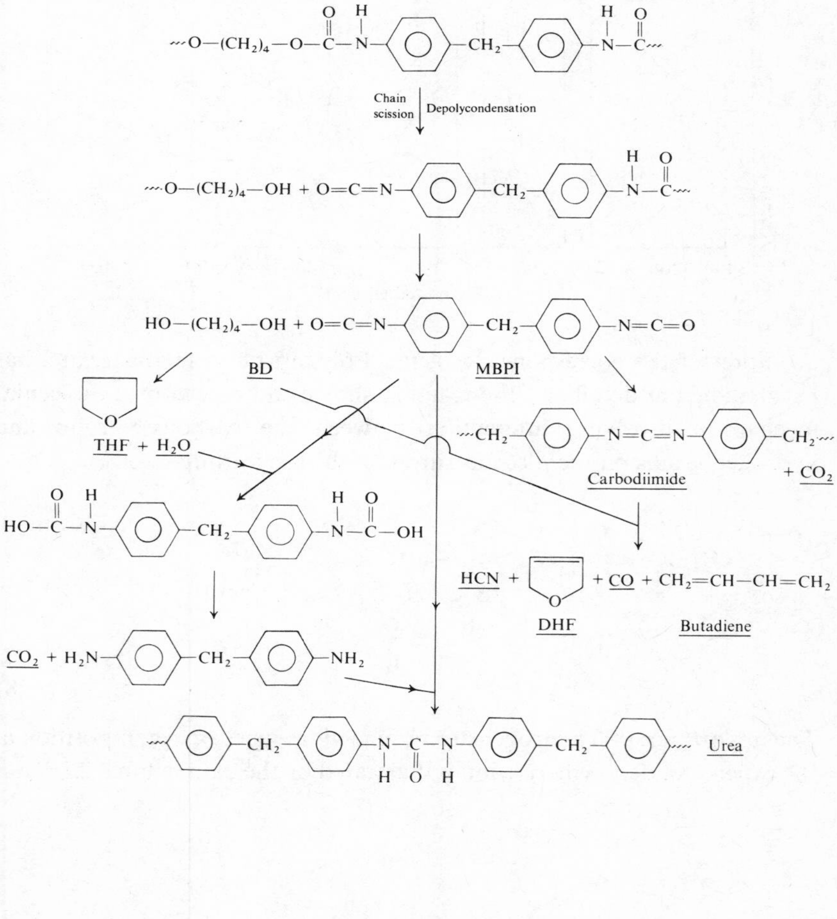

2.4. Substituent reactions

Polymethacrylates

Although, as described in an earlier section of this chapter, poly(methyl methacrylate) degrades thermally in a radical process to give quantitative yields of monomer, in higher polymethacrylates decomposition of the ester group may occur, resulting in methacrylic acid units in the polymer and the

Figure 2.11. Infra-red spectra of a polyurethane heated to various temperatures at $10°$ min^{-1}. (*a*) original polymer; (*b*) 300 °C; (*c*) 350 °C; (*d*) 450 °C. (Reproduced by kind permission of *J. Polym. Sci. Chem. Ed.*, **16**, 1563, 1978.)

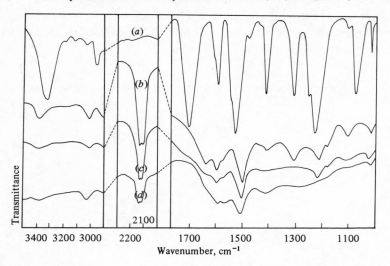

evolution of the corresponding olefin. Poly(tert-butyl methacrylate) has been studied in detail and the reaction shown to proceed by a molecular mechanism involving interaction between the carbonyl group and hydrogen atoms on the β carbon atom of the ester group.

$$\text{(2.28)}$$

The importance of β hydrogen atoms in facilitating ester decomposition at the expense of depolymerisation is illustrated by the data in table 2.2. Thus ester decomposition only becomes important when the monomer unit incorporates at least five hydrogen atoms on the β carbon and depolymerisation is quantitative when there are at most one or two β hydrogen atoms.

The vital factor which determines whether depolymerisation or ester decomposition will predominate is the relative facility of the latter. If a significant proportion of the ester groups is destroyed during the early stages of heating then the residual methacrylic acid units (or methacrylic anhydride units formed by elimination of water) block the unzipping process and thus inhibit formation of monomer. If the radical depolymerisation reaction can be initiated at a lower temperature than ester decomposition, for example by UV radiation (see chapter 4), then ester decomposition, even in poly(tert-butyl methacrylate) is replaced by quantitative production of monomer.

Poly(vinyl acetate)

Ester decomposition also occurs in poly(vinyl esters) but in this case carboxylic acid is liberated and olefinic double bonds appear in the polymer chain backbone, reaction (2.29). For example in the case of poly(vinyl acetate),

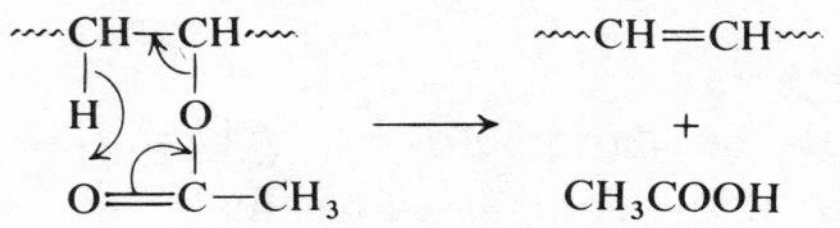

$$\text{(2.29)}$$

the β hydrogen atom is effectively interacting as a proton with the oxygen atom, so that the reaction should be facilitated by electron-attracting groups in the vicinity. It is not surprising, therefore, that the electron

Table 2.2. *Mechanisms of thermal degradation of poly(alkyl methacrylates). Numbers indicate the number of hydrogen atoms on the β-carbon atom.*

Depolymerisation		Mainly depolymerisation		Ester decomposition	
methyl	0	ethyl	2	iso-propyl	6
neo-pentyl	0	n-propyl	2	sec-butyl	5
iso-butyl	1	n-butyl	2	tert-butyl	9
ethoxyethyl	2	n-hexyl	2		
		n-heptyl	2		
		n-octyl	2		

attracting properties of the carbon–carbon double bond causes the reaction in poly(vinyl acetate) to pass from unit to unit along the chain by reaction (2.30),

$$
\sim\text{CH}=\text{CH}-\overset{\delta-}{\text{CH}}-\text{CH}\sim \qquad \sim\text{CH}=\text{CH}-\text{CH}=\text{CH}\sim
$$

$$
{}^{\delta+}\text{H} \quad \text{O} \qquad\qquad +
$$

$$
\text{O}=\text{C}-\text{CH}_3 \qquad\qquad \text{CH}_3\text{COOH} \qquad (2.30)
$$

the ultimate effect being to produce extended conjugation and colour. The overall process may therefore be regarded as a non-radical chain reaction in which initiation consists of the loss of an acid molecule from a saturated segment of the chain and propagation consists of the double bond activated loss of acid. This molecular process should be expected to pass relatively slowly along the chain but there is no obvious termination process so that conjugated sequences of considerable length should be built up and the number of degrading sequences should increase continuously during the reaction. It is these features of the reaction which cause the colour of the residual polymer to pass relatively slowly through the spectrum from yellow through orange to red and the evolution of acetic acid to exhibit autocatalytic properties.

Poly(vinyl chloride)

The products of thermal degradation of poly(vinyl chloride) are hydrogen chloride and a coloured, highly-conjugated residue

$$
\sim\text{CH}_2\text{-CHCl-CH}_2\text{-CHCl}\sim \rightarrow \sim\text{CH}=\text{CH-CH}=\text{CH}\sim \quad +\text{HCl}
$$

$$
(2.31)
$$

which are analogous to those from poly(vinyl acetate). Reaction (2.31) occurs so readily that it has been said that if poly(vinyl chloride) had not been discovered until the present time it would have been discarded after preliminary assessment as unsuitable for commercial development in competition with existing materials. It was, however, one of the first of the modern high-tonnage plastics to be developed and most of the major stability problems had been solved as a result of the discovery of very efficient stabilisers. Nevertheless the mechanism of degradation of poly(vinyl chloride) is still not completely understood in spite of the fact that more research time has probably been devoted to this polymer than to any other.

The rate of formation of HCl from poly(vinyl chloride) is very dependent upon experimental conditions and it is clear that much of the disagreement which exists about the reaction mechanism is due to the fact that the preparation of the samples and the conditions under which the degradation experiment was carried out have not been precisely defined. For example with very small fine powder samples, the formation of HCl is linear with time. The reaction becomes autocatalytic however with larger samples, with large particles rather than powder or if HCl is allowed to accumulate in the gas phase above the degrading polymer. These observations have been interpreted as being due to a catalytic effect by HCl.

The rate of formation of HCl is accelerated by oxygen and under certain conditions by radical producing compounds like azo-bis-isobutyronitrile. It has also been claimed that hydroquinone, a radical reaction inhibitor, slows down the reaction, all of which strongly indicates a radical mechanism.

There is a great deal of disagreement about the effect of molecular weight on the stability of poly(vinyl chloride). Some investigators claim that there is no effect while others say that the rate of evolution of HCl is inversely proportional to the molecular weight, that is, proportional to the number of chain ends, which suggests that initiation occurs at the chain ends. Another group has shown that while there is a variation in rate from sample to sample, it does not vary regularly with molecular weight. For reasons of this kind it has been concluded that initiation of HCl evolution must be associated with the presence of more than one type of 'weak link' in the polymer chain.

While HCl is liberated from poly(vinyl chloride), new absorption bands appear in the UV and visible spectrum. These are associated with the polyene structures formed which result in discolouration of the polymer. The position of these absorption bands suggests that the polyene sequence lengths do not much exceed ten. It is also clear that, unlike the situation in degrading poly(vinyl acetate), once initiated, these sequences develop very rapidly as one might expect in a radical process.

It is clear that the temperatures at which HCl is liberated from poly(vinyl chloride) are very much lower than might be expected by comparison with the behaviour of small molecular 1,3-chloro-substituted model compounds. This led to the proposal that the reaction is initiated at more labile structures present in relatively low concentration. To investigate this a large number of model compounds have been studied and relative orders of stability and approximate decomposition temperature are as follows:

$$CH_3-CH=\overset{\underset{\textstyle |}{Cl}}{C}-CH_2-CH_3 \ggg CH_3-\overset{\underset{\textstyle |}{Cl}}{CH}-CH_2-\overset{\underset{\textstyle |}{Cl}}{CH}-CH_3 > CH_3-\overset{\underset{\textstyle |}{Cl}}{CH}-CH_3$$

$$400\,°C \qquad\qquad 360\,°C \qquad\qquad 340\,°C$$

$$> CH_2=CH-CH_2-\overset{\underset{\textstyle |}{Cl}}{CH}-CH_2-CH_3 > CH_2=CH-\overset{\underset{\textstyle |}{Cl}}{CH}-CH_2-CH_3$$

$$325\,°C \qquad\qquad\qquad 280\,°C$$

$$> CH_3-\overset{\overset{\textstyle CH_3}{\textstyle |}}{\underset{\underset{\textstyle Cl}{\textstyle |}}{C}}-CH_3 > CH_3-CH_2-\overset{\overset{\textstyle C_2H_5}{\textstyle |}}{\underset{\underset{\textstyle Cl}{\textstyle |}}{C}}-CH_2-CH_3$$

$$230\,°C \qquad\qquad\quad 180\,°C$$

$$> CH_3-CH=CH-\overset{\underset{\textstyle |}{Cl}}{CH}-CH_2-CH_3$$

$$160\,°C$$

Thus, although the instability of poly(vinyl chloride) has not been attributed to any single labile abnormality, structures incorporating tertiary or allylic chlorine atoms must be particularly suspect and may easily be formed during the polymerisation process or during subsequent storage or processing of the polymer.

Both free radical and molecular mechanisms have been proposed for the thermal dehydrochlorination of poly(vinyl chloride). The various radical mechanisms suggested are summarised in reactions (2.32)–(2.37). Initiation involves liberation of a chlorine atom from a labile centre;

$$\text{\textasciitilde\textasciitilde CH}_2-\overset{\textstyle |}{\underset{\underset{\textstyle Cl}{\textstyle |}}{C}}\text{\textasciitilde\textasciitilde} \longrightarrow \text{\textasciitilde\textasciitilde CH}_2-\overset{\textstyle |}{C}\text{\textasciitilde\textasciitilde} + Cl\cdot \quad \text{Initiation} \qquad (2.32)$$

$$Cl\cdot + \text{\textasciitilde\textasciitilde}CH_2-CHCl-CH_2-CHCl\text{\textasciitilde\textasciitilde} \qquad\qquad (2.33)$$

$$\longrightarrow HCl + \text{\textasciitilde\textasciitilde}\overset{\textstyle |}{C}H-CHCl-CH_2-CHCl\text{\textasciitilde\textasciitilde} \qquad (2.34)$$

$$\longrightarrow \text{\textasciitilde\textasciitilde}CH=CH-CH_2-CHCl\text{\textasciitilde\textasciitilde} + Cl\cdot \quad\rbrace\ \text{Propagation} \qquad (2.35)$$

$$\longrightarrow \text{\textasciitilde\textasciitilde}CH=CH-\overset{\textstyle |}{C}H-CHCl\text{\textasciitilde\textasciitilde} + HCl \qquad (2.36)$$

$$\longrightarrow \text{\textasciitilde\textasciitilde}CH=CH-CH=CH\text{\textasciitilde\textasciitilde} + Cl\cdot \qquad\qquad (2.37)$$

The chlorine atoms are believed to react with the most convenient hydrogen atoms, namely the adjacent methylene hydrogen atoms thus leading to conjugated unsaturation. Occasionally they may escape from this environment and attack another chain;

$$Cl\cdot + \sim CH_2\text{--}CHCl\text{--}CH_2\text{--}CHCl\sim \rightarrow HCl + \sim CH_2\text{--}CHCl\text{--}\dot{C}H\text{--}CHCl\sim$$

$$(2.38)$$

This would account for the low average length of polyene sequences revealed by UV spectra. The reaction is terminated by reaction (2.39),

$$Cl\cdot + Cl\cdot \rightarrow Cl_2 \qquad (2.39)$$

dimerisation of polymer radicals being unlikely because of their low mobility in the degrading polymer.

Many experimental observations cannot be explained by a radical mechanism, and the effect of HCl is particularly difficult to explain. However, molecular elimination of HCl can be envisaged as proceeding through a cyclic transition state, reaction (2.40),

$$\sim CH_2\text{--}\underset{\underset{Cl}{|}}{CH}\text{--}\underset{\underset{H}{|}}{CH}\text{--}CH\text{--}CHCl\sim \longrightarrow HCl + \sim CH_2\text{--}CH{=}CH\text{--}CHCl\sim$$

$$(2.40)$$

and a modified process, reaction (2.41), seems possible in presence of HCl.

$$\sim CH_2\text{--}\underset{\underset{Cl}{\diagdown}}{CH}\text{--}\overset{\overset{H\text{--}Cl}{}}{\underset{\underset{H}{\diagup}}{CH}}\text{--}CHCl\sim \longrightarrow \sim CH\text{--}CH{=}CH\text{--}CHCl\sim + 2HCl.$$

The chlorine atom on the next carbonyl atom will be activated by the unsaturation so the reaction will tend to pass from unit to unit along the chain to produce conjugation. The situation in the presence of oxygen is different since it leads to peroxide catalysts. This will be discussed in chapter 4. Thus the current view is that although the radical process is probably the more important, both radical and molecular reactions can occur. Which one predominates depends upon many factors such as polymer preparation, the presence of unsaturated impurities, sample form, degradation temperature and the presence of oxygen-containing impurities.

2.4.1. Cyclization reactions with elimination

Cyclization reactions frequently occur between adjacent substituents on polymer molecules, usually with the elimination of a small molecule. For example, poly(N-methyl methacrylamide) eliminates methylamine, reaction (2.42),

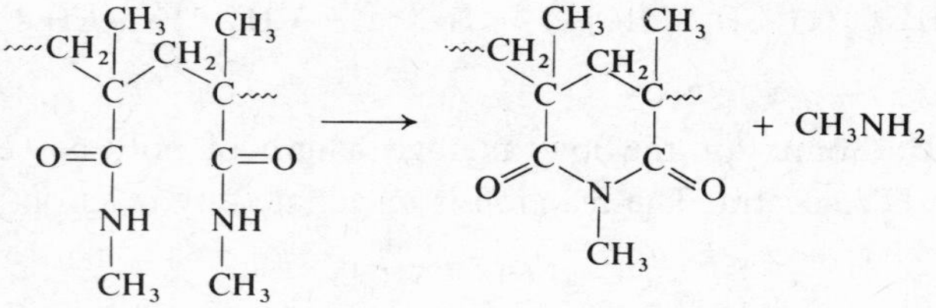

(2.42)

whilst poly(methyl vinyl ketone), poly(methyl isopropenyl ketone)

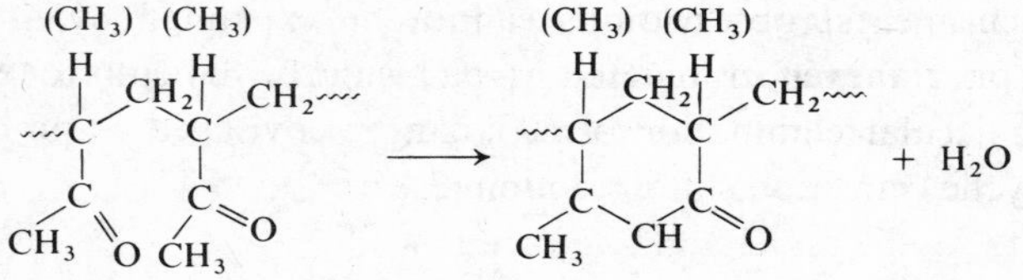

(2.43)

and poly(methacrylic acid) eliminate water (reactions (2.43) and (2.44)).

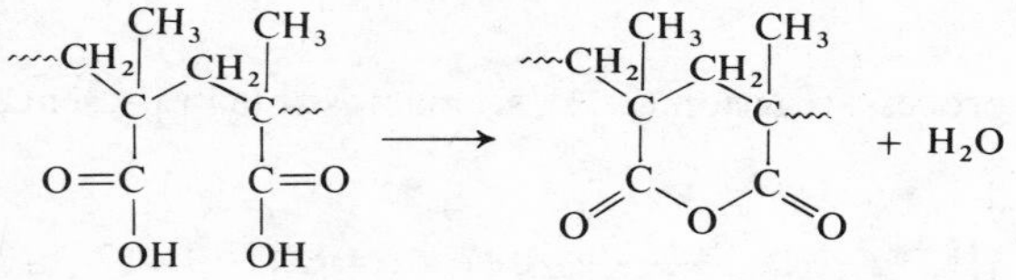

(2.44)

Although corresponding intermolecular reactions should also be possible to form cross-linked structures, intramolecular cyclization processes invariably predominate, presumably because the regular distribution of substituents on alternate carbon atoms in predominantly head to tail polymers sterically favours intramolecular interaction.

Poly(methacrylic acid) has been mentioned earlier in this chapter as the primary product of decomposition of certain methacrylates and provides an interesting illustration of the application of IR spectroscopy to the investigation of polymer degradation reactions. Spectra of poly(methacrylic acid) and its degradation product are shown in Fig. 2.12 which demonstrates conversion of acid to anhydride. Thus broad absorption at 3600–2500 cm^{-1}, which is characteristic of acid hydroxyl, is replaced by intense C–O–C absorption at 1022 cm^{-1} and splitting of the carbonyl peak (1795 and 1750 cm^{-1}) both of which are characteristic of anhydride. Closer observation of the nature of the carbonyl splitting makes it possible to deduce that the water has been eliminated intramolecularly to form cyclic structures rather than intermolecularly to form cross links.

Polyacrylonitrile

When polyacrylonitrile was being developed as a textile fibre about 25 years ago, problems of discolouration in use were clearly due to degradation. These problems were solved through studies of the mechanism of the discolouration process. About 10–15 years ago a second phase of interest in the thermal degradation of polyacrylonitrile was initiated by the discovery that very strong fibres, useful for the reinforcement of synthetic polymers as structural materials, could be prepared by heating polyacrylonitrile through a carefully controlled temperature programme to temperatures in excess of 1000 °C. The production of 'carbon fibres' obviously involved a degradation process and to produce carbon fibres with optimum physical properties this degradation had to be understood and controlled.

The most important features of the thermal degradation of polyacrylonitrile, from the point of view of reaction mechanism, are as follows.

1. Colouration occurs in polyacrylonitrile at about 250 °C and is due to the conjugation resulting from polymerisation or oligomerisation of sequences of adjacent nitrile groups.

Figure 2.12. Infra-red spectra of A poly(methacrylic acid) and B its residue after heating at 200 °C. (Reproduced by kind permission of *Polymer Science*, ed. A. D. Jenkins, North Holland Publishing Company, 1972, p. 1494.)

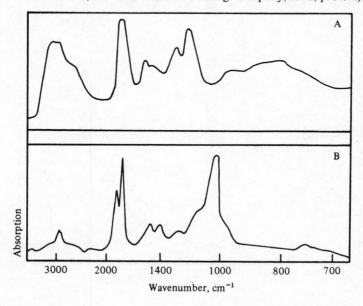

$$\text{(2.45)}$$

2. Colouration involves a single spectral colour, intensifying from light to dark brown. It does not pass through the spectrum from yellow to red as it would if conjugation developed slowly.

3. The most outstanding feature of the reaction is the sudden evolution of a large amount of heat which may even cause the degrading sample to explode. This is illustrated by the DSC curves in Fig. 2.13. The suddenness and intensity of the liberation of heat is demonstrated not only by the height of the exotherm but also by the tilt which is a measure of the temperature difference between the sample and the reference cell.

4. The sharp initial peak in the TVA curve (Fig. 2.14) is associated with the exotherm and is due to ammonia and hydrogen cyanide. Products non-condensible at $-196\,°C$ only appear above $350\,°C$. This is hydrogen

Figure 2.13. DSC curves for polyacrylonitrile (10 mg) heated under flowing nitrogen $(1\,l\,min^{-1})$ at 1, $10\,°C\,min^{-1}$ and 2, $5\,°C\,min^{-1}$. (Reproduced by kind permission of *Developments in Polymer Degradation – 1*, ed. N. Grassie, App. Sci. Pub., London, 1977, p. 141.)

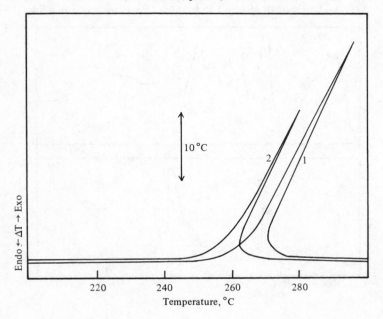

corresponding to aromatisation of the cyclic structure. Nitrogen appears at 850–900 °C and the residue becomes almost pure graphitic carbon.

5. The TG curve in Fig. 2.15 shows a sudden loss in weight near the exotherm. This cannot be accounted for by the ammonia and hydrogen cyanide but proves to be chain fragments similar to residual polymer.

Through detailed investigation of these features of the reaction the following facts have been established:

Figure 2.14. TVA curves for polyacrylonitrile. Heating rate 10°C min⁻¹. Trap temperatures: —— 0°C; ····· −75°C; — — − 196°C. (Reproduced by kind permission of *Developments in Polymer Degradation – 1*, ed. N. Grassie, App. Sci. Pub., London, 1977, p. 143.)

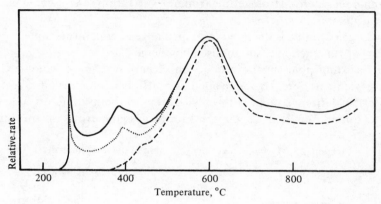

Figure 2.15. TG curve for polyacrylonitrile (10 mg). Heating rate 10°C min⁻¹ under 80 ml min⁻¹ nitrogen flow. (Reproduced by kind permission of *Developments in Polymer Degradation – 1*, ed. N. Grassie, App. Sci. Pub., London, 1977, p. 142.)

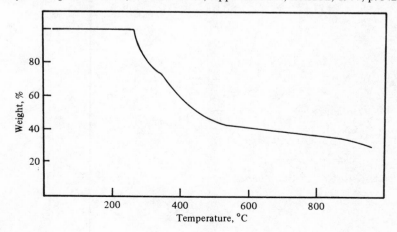

(*a*) it is a radical process;

(*b*) the large amount of heat evolved is due to the exothermicity of the cyclization reaction;

(*c*) in the absence of impurities in the polymer, initiation occurs at labile structural abnormalities;

(*d*) the length of conjugated nitrile sequences (the 'zip' length) is short but;

(*e*) the kinetic chain length is long, being maintained by transfer processes;

(*f*) this produces a polymer chain consisting of short cyclized segments linked by segments of unchanged monomer units;

(*g*) chain scission occurs at unchanged segments to produce chain fragments (cold ring fraction);

(*h*) ammonia is derived from terminal imine (N–H) groups in the conjugated sequences;

(*i*) hydrogen cyanide is liberated from unchanged acrylonitrile units.

The overall reaction is summarised in scheme 2.2.

The starting point in the production of carbon fibres is commercial polyacrylonitrile fibre. To obtain good carbon fibre it is important to retain this original fibre structure throughout the carbonisation process to temperatures over 1000 °C. The great problem is the exotherm since the

Scheme 2.2 Mechanism of the cyclisation of polyacrylonitrile.

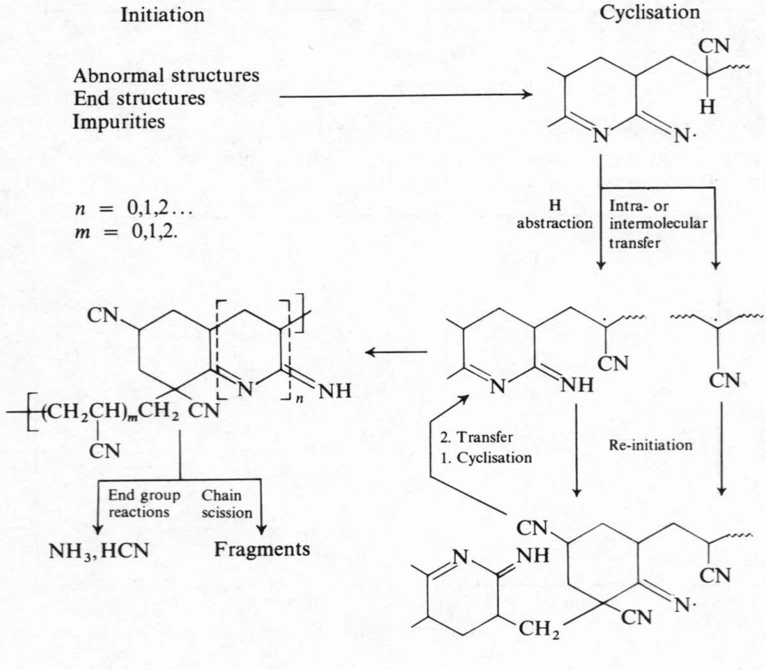

intense heat generated tends to destroy the fibre structure. For this reason the fibre must be heated in air for several hours at 200 °C. Oxygen appears to fulfil a dual function. In the first place it initiates the reaction at a lower temperature, thus reducing the intensity of the exothermic reaction. More importantly, however, it redirects the course of the reaction by initially forming allylic hydroperoxide groups which then decompose to give carbonyl (see scheme 2.3). Hydrogen bonded interaction between carbonyl groups and secondary amine groups in neighbouring polymer chains results in effective physical stabilisation of the polymer structure, which allows the fibres to maintain their physical integrity during subsequent degradation reactions.

It has been demonstrated that the time consuming and therefore expensive preoxidation process may be circumvented by similarly broadening the exotherm by the use of cyclization initiating impurities or comonomers.

Polymethacrylonitrile (X) was referred to (table 2.1) as a material which, like poly(methyl methacrylate) degrades quantitatively to monomer. This is only strictly true if the polymer is free from certain types of impurities. For example, in presence of traces of acidic or basic impurities or methacrylic acid as comonomer, cyclization occurs as in polyacrylonitrile. This,

Scheme 2.3 Physical stabilisation of PAN in the presence of oxygen.

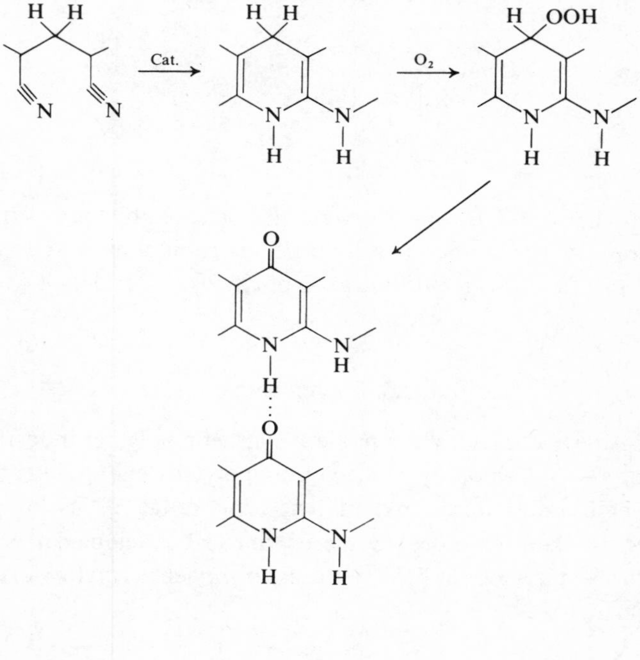

however, appears to be an ionic type reaction which passes relatively slowly from unit to unit along the polymer chain (scheme 2.4) so that, unlike the single spectral colour in polyacrylonitrile, colour develops slowly through the spectrum from yellow through orange to red and black.

Scheme 2.4. Cyclisation of polymethacrylonitrile.

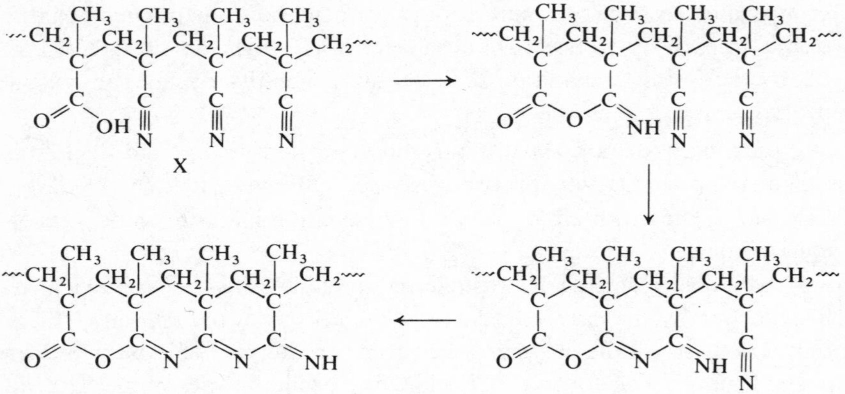

A similar reaction is induced by alkali at ambient temperatures in solutions of both polyacrylonitrile and polymethacrylonitrile.

Analogous cyclization reactions do not occur in poly(α-chloro-acrylonitrile(XXIII)).

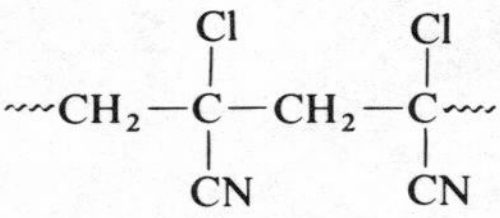

XXIII

In this polymer the C–Cl bond is so much weakened by the neighbouring nitrile group that C–Cl bond scission occurs rapidly even at 140 °C and initiates loss of hydrogen chloride as in poly(vinyl chloride).

2.5. Copolymers

In the early development of the modern synthetic polymer industry, some 30–40 years ago, homopolymers, such as polyethylene, polystyrene and poly(vinyl chloride) were predominent. The range of useful physical properties was later extended by the commercial development of copolymers such as styrene/butadiene, ethylene/propylene, vinyl chloride/vinyl acetate and many others. However the introduction of a comonomer into a homopolymer molecule may profoundly affect its stability and degradation properties.

Styrene/α-chloroacrylonitrile and methyl methacrylate/α-chloroacrylonitrile

A small proportion of copolymerised α-chloroacrylonitrile is a very effective destabiliser for a number of homopolymers and its effects on poly(methyl methacrylate) and polystyrene are particularly well understood.

No significant quantities of volatile products are formed from poly(methyl methacrylate) below 200 °C. When α-chloroacrylonitrile units are copolymerised into the molecules, however, volatile products, methyl methacrylate, hydrogen chloride and a trace of α-chloroacrylonitrile, appear even at 140 °C. Concurrently, there is a rapid decrease in molecular weight as shown in Fig. 2.16.

Figure 2.16. Changes in molecular weight with time of heating at various temperatures for a copolymer of methyl methacrylate and α-chloroacrylonitrile (molar ratio 94/6). (Reproduced by kind permission of *Europ. Polym. J.*, **2**, 255, 1966.)

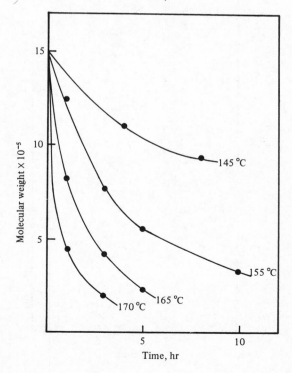

When chain scission and volatilisation occur together, the chain length at time t, P_t, is given by

$$P_t = \frac{P_0(1-x)}{s+1} \qquad (vi)$$

in which P_0 is the initial chain length, x is the fraction of polymer volatilised and s is the average number of bonds broken per chain. Thus the number of bonds broken per monomer unit s/P_0 is given by

$$s/P_0 = \frac{1-x}{P_t} - \frac{1}{P_0}. \qquad (vii)$$

The nature of the chain scission at α-chloroacrylonitrile units is illustrated by the linear plots in Fig. 2.17. The production of monomer is also linear with time so that for each chain scission a fixed amount of

Figure 2.17. Chain scission plot of the data in Fig. 2.16. (Reproduced by kind permission of *Europ. Polym. J.*, **2**, 255, 1966.)

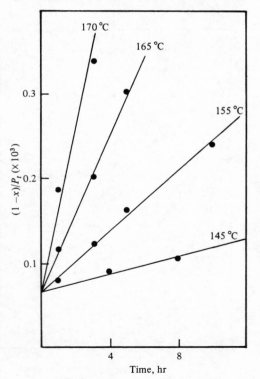

Time, hr

monomer is formed and this is strong evidence that chain scission results in radicals which depolymerise to monomer.

A similar random scission reaction also occurs at 140 °C in copolymers of styrene and α-chloroacrylonitrile. In this case, however, HCl is the only volatile product and styrene only appears at 280 °C as in the case of pure polystyrene.

Since degradation in these two copolymers starts at 140 °C as in the homopolymer of α-chloroacrylonitrile (discussed in the previous section) it seems that the same initiation process is involved. Thus in the styrene copolymer, chain scission and the reaction products may be accounted for as follows:

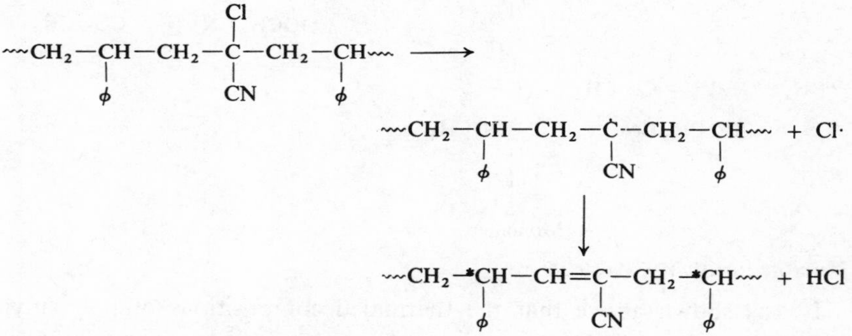

$$\text{\small{$\sim$CH}}_2\text{—CH—CH}_2\text{—C—CH}_2\text{—CH}\sim \longrightarrow$$

(with Cl and CN substituents, φ groups)

$$\sim\text{CH}_2\text{—CH—CH}_2\text{—}\dot{\text{C}}\text{—CH}_2\text{—CH}\sim + \text{Cl}\cdot$$

$$\downarrow$$

$$\sim\text{CH}_2\text{—}^*\text{CH—CH}=\text{C—CH}_2\text{—}^*\text{CH}\sim + \text{HCl}$$

$$(2.46)$$

The bonds indicated by an asterisk, being activated by both the ethylenic unsaturation and the aromatic ring, could be reduced in strength by as much as 145–165 kJ mole^{-1} compared with normal C–C bonds. Bond scission apparently occurs, followed by the formation of stable molecules by hydrogen transfer;

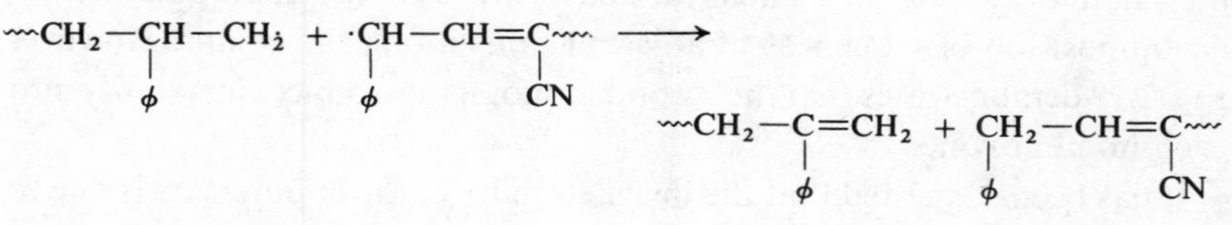

$$\sim\text{CH}_2\text{—CH—CH}_2 + \cdot\text{CH—CH}=\text{C}\sim \longrightarrow$$

$$\sim\text{CH}_2\text{—C}=\text{CH}_2 + \text{CH}_2\text{—CH}=\text{C}\sim$$

$$(2.47)$$

In the methyl methacrylate system the reaction may be similarly interpreted until the last step when the radicals are apparently capable of depolymerising to monomer, as shown in scheme 2.5:

Scheme 2.5

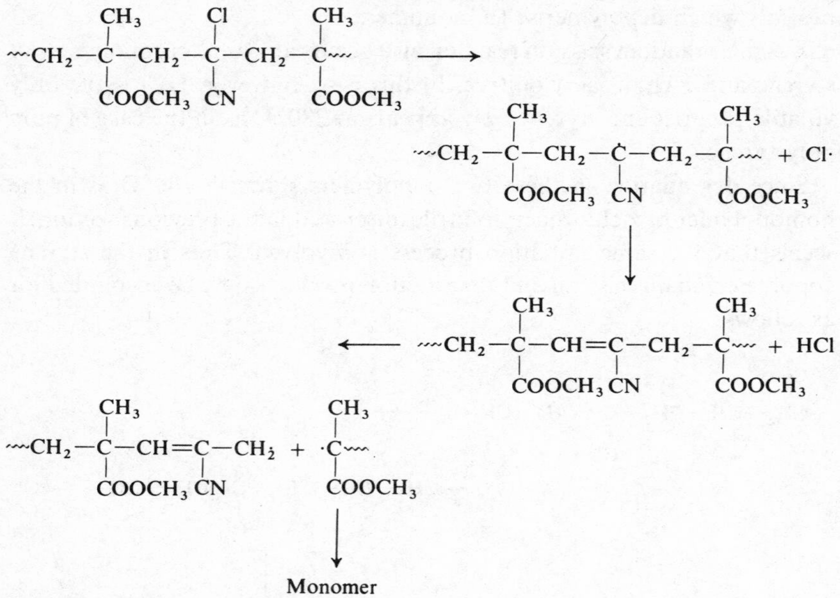

$$\text{~CH}_2-\overset{\overset{\displaystyle CH_3}{|}}{\underset{\underset{\displaystyle COOCH_3}{|}}{C}}-CH_2-\overset{\overset{\displaystyle Cl}{|}}{\underset{\underset{\displaystyle CN}{|}}{C}}-CH_2-\overset{\overset{\displaystyle CH_3}{|}}{\underset{\underset{\displaystyle COOCH_3}{|}}{C}}\text{~} \longrightarrow$$

$$\text{~CH}_2-\overset{\overset{\displaystyle CH_3}{|}}{\underset{\underset{\displaystyle COOCH_3}{|}}{C}}-CH_2-\overset{\cdot}{\underset{\underset{\displaystyle CN}{|}}{C}}-CH_2-\overset{\overset{\displaystyle CH_3}{|}}{\underset{\underset{\displaystyle COOCH_3}{|}}{C}}\text{~} + Cl\cdot$$

$$\downarrow$$

$$\longleftarrow \quad \text{~CH}_2-\overset{\overset{\displaystyle CH_3}{|}}{\underset{\underset{\displaystyle COOCH_3}{|}}{C}}-CH=C-CH_2-\overset{\overset{\displaystyle CH_3}{|}}{\underset{\underset{\displaystyle COOCH_3}{|}}{C}}\text{~} + HCl$$
(with CN on middle carbon)

$$\text{~CH}_2-\overset{\overset{\displaystyle CH_3}{|}}{\underset{\underset{\displaystyle COOCH_3}{|}}{C}}-CH=C-CH_2^{\cdot} + \overset{\overset{\displaystyle CH_3}{|}}{\underset{\underset{\displaystyle COOCH_3}{|}}{\cdot C}}\text{~}$$
(with CN on the =C)

$$\downarrow$$

Monomer

Vinyl acetate/vinyl chloride

It was shown above that the thermal decompositions of poly(vinyl chloride) and poly(vinyl acetate) are similar in the sense that acid is liberated and the same conjugated residue remains. The mechanisms may be different – radical for poly(vinyl chloride) and molecular for poly(vinyl acetate) – but conjugation and colour appear early in the reaction so it is clear that in each case unsaturation facilitates the decomposition of adjacent units.

In the copolymer, colouration occurs as in the homopolymers and the two acids are produced in the proportions in which the monomers are present in the copolymer. Thus reaction occurs as in the homopolymers by decomposition of sequences of adjacent units along the chains. However Fig. 2.18 demonstrates that the second monomer introduces instability into each homopolymer.

It has been suggested that the lower stability of the copolymers is due to the influence of vinyl chloride units on adjacent vinyl acetate units.

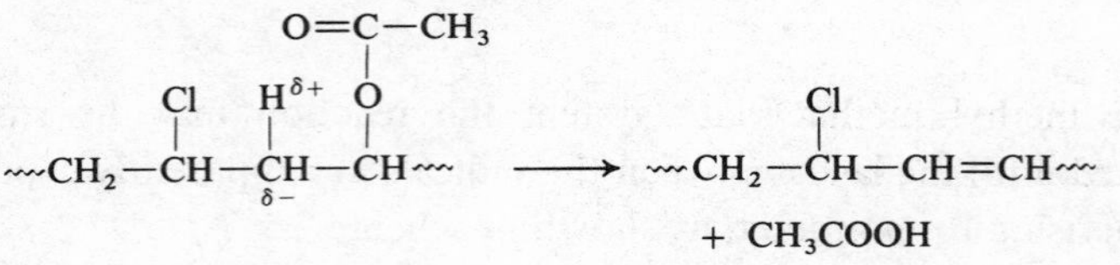

$$\text{~CH}_2-\underset{\underset{\displaystyle Cl}{|}}{CH}-\underset{\delta-}{\overset{\overset{\displaystyle H^{\delta+}}{|}}{CH}}-\overset{\overset{\displaystyle O=C-CH_3}{\underset{\displaystyle O}{|}}}{CH}\text{~} \longrightarrow \text{~CH}_2-\underset{\underset{\displaystyle Cl}{|}}{CH}-CH=CH\text{~}$$

$$+ CH_3COOH \qquad (2.48)$$

Thus the electron withdrawing power of the chlorine atom will, like that of the double bond, activate the adjacent methylene group and facilitate the molecular decomposition mechanism by which acetic acid is liberated. Once this first double bond is formed it will activate decomposition of adjacent vinyl acetate or vinyl chloride units.

Methyl methacrylate/alkyl acrylate

A number of comonomers whose homopolymers do not depolymerise to monomer are capable of blocking monomer-producing depolymerisation reactions. For example, a number of alkyl acrylates and acrylonitrile behave in this way in poly(methyl methacrylate) and small amounts of copolymerised ethyl acrylate have been used industrially to stabilise poly(methyl methacrylate) toward depolymerisation during processing. Depolymerisation proceeds through the methyl methacrylate units from the point of initiation but acrylate units cannot be liberated so that the terminal acrylate radicals are instead deactivated.

Methyl methacrylate/styrene

Polystyrene is much more thermally stable than poly(methyl methacrylate) and their degradation rate curves are quite different in shape. Rate curves at

Figure 2.18. Influence of copolymer composition on the rate of evolution of acid from copolymers of vinyl acetate and vinyl chloride at 180 °C. (Reproduced by kind permission of *Polymer Science*, ed. A. D. Jenkins, North Holland Publishing Company, 1972, p. 1502.)

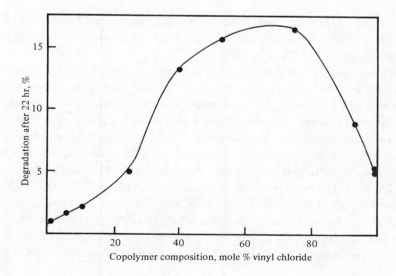

various temperatures for a series of polymers and copolymers are illustrated in Fig. 2.19. The increase in stability with styrene content is clear from the temperatures required and the transition from poly(methyl methacrylate) to polystyrene type behaviour is also obvious. The effect of styrene units on the stability of poly(methyl methacrylate) is disproportionately large. For example, the initial rate of depolymerisation of the 1/4, styrene/methyl methacrylate copolymer degraded at 300 °C is roughly half of that of pure poly(methyl methacrylate) degraded at 260 °C. If, as a rough guide, it is assumed that the rate doubles for a 10° rise in temperature, then the 1/4 copolymer is 30–40 times more stable than poly(methyl methacrylate).

The reasons for this can be found in the nature of the chain ends. Approximately 50% of poly(methyl methacrylate) molecules have unsaturated degradable ends formed by disproportionation. But it has been demonstrated that in a copolymerising mixture of styrene and methyl methacrylate, most termination occurs between unlike radicals, even when the proportion of styrene is quite low, and that this 'cross' termination is almost exclusively combination by reaction (2.49).

$$\text{\textasciitilde\textasciitilde CH}_2-\underset{\underset{\text{COOCH}_3}{|}}{\overset{\overset{\text{CH}_3}{|}}{\text{C}}}\cdot \quad + \quad \cdot\underset{\underset{\phi}{|}}{\text{CH}}-\text{CH}_2\text{\textasciitilde\textasciitilde} \quad \longrightarrow \quad \text{\textasciitilde\textasciitilde CH}_2-\underset{\underset{\text{COOCH}_3}{|}}{\overset{\overset{\text{CH}_3}{|}}{\text{C}}}-\underset{\underset{\phi}{|}}{\overset{\overset{\phi}{|}}{\text{CH}}}-\text{CH}_2\text{\textasciitilde\textasciitilde} \qquad (2.49)$$

Thus even 1/4, styrene/methyl methacrylate copolymer contains only a small proportion of unsaturated terminal structures through which degradation is most readily initiated.

One may therefore summarise copolymer stability data as follows. A comonomer may make a homopolymer unstable for two general reasons. First, it may act as an initiator of the degradation reaction which normally occurs in the homopolymer as in the vinyl chloride/vinyl acetate system, for example. Secondly, the comonomer may have an unstable structure which can induce a new kind of degradation reaction as in the effect of α-chloroacrylonitrile on polystyrene and poly(methyl methacrylate).

On the other hand, stability is usually conferred by a comonomer if it interferes with the course of a degradation process as the alkyl acrylates block depolymerisation in poly(methyl methacrylate). However the increased stability which styrene confers on poly(methyl methacrylate) can not be explained in this way. It is really an unexpected bonus which arises from the fact that the presence of styrene in the polymerising mixture modifies the mechanism of polymerisation in a significant way.

2.6. Polymer blends

Following copolymerisation, the blending, or mixing, of polymers is becoming increasingly important as a method of modifying and extending polymer properties to satisfy commercial requirements. The blending of polystyrene with an elastomer such as polyisoprene or a copolymer of butadiene and acrylonitrile to form the 'impact' polystyrenes with none of the brittleness of polystyrene, is an outstanding example.

Structure and preparation of blends

Whereas polymers and random copolymers are homogeneous materials, blends are heterogeneous with a dispersed phase of one polymer in a

Figure 2.19. Rate curves for the volatilisation of styrene/methyl methacrylate copolymers and the homopolymers. (Reproduced by kind permission of *Polymer Science*, ed. A. D. Jenkins, North Holland Publishing Company, 1972, p. 1479.)

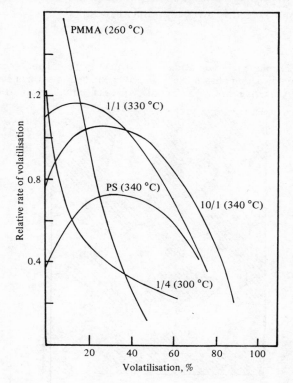

continuous phase of the other as illustrated schematically in Fig. 2.20. The domain size will normally be in the range 1–10 μ. It is to be anticipated that degradation processes characteristic of each constituent polymer will proceed in the appropriate phase. Any interactions between these processes would have to take place across phase boundaries either by direct interaction of macromolecules or macroradicals at the boundary or, more likely, by diffusion of a small molecule or radical, formed in one polymer, across the boundary to react with the other polymer.

Because of the possible significance of the phase boundary, and thus of domain size and distribution in blends, it is important to give particular attention to their preparation for quantitative degradation investigations. They may be prepared in a number of ways as follows: (*a*) as a film by evaporation of a solution of the two polymers; (*b*) as a powder coprecipitated from a solution of the two polymers; (*c*) by freeze drying a mixture of the polymers; (*d*) by grinding together the polymers as powders; (*e*) by melt processing. Each method has its advantages and disadvantages. The last is the normal industrial method but (*a*) and (*d*) are most useful for small scale laboratory work. However an important ultimate test of the method is whether or not it readily gives reproducible results.

Figure 2.20. Schematic representation of the structure of a polymer blend showing continuous and dispersed phases. (Reproduced by kind permission of *Developments in Polymer Degradation – 1*, ed. N. Grassie, App. Sci. Pub., London, 1977, p. 173.)

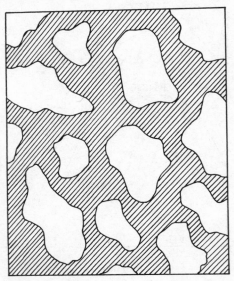

Degradation technique

The experimental techniques which have been outlined earlier in this chapter have been equally successfully applied to polymer blends. One very simple procedure which has been found to be especially effective in thermal analysis work for the detection of interaction between the constituents of a blend is to make a close comparison of degradations of the two polymers separately but simultaneously and of the same amounts of the two polymers blended together. In TVA, for example, these are carried out in twin-limbed degradation tubes which are illustrated diagrammatically in Fig. 2.21. These two situations will be referred to as 'unmixed' and 'mixed' respectively and examples will emerge in the discussion which follows.

The thermal degradation of a large number of blend pairs has been studied in considerable detail. The principal features will be illustrated by reference to two in which poly(vinyl chloride) is one of the constituents.

Poly(vinyl chloride) – poly(methyl methacrylate)

Mixed and unmixed TVA curves for a 50/50, w/w blend of poly(vinyl chloride) and poly(methyl methacrylate) are shown in Fig. 2.22. It should be noted that a poly(methyl methacrylate) of very high molecular weight was used so that the initial peak due to initiation at unsaturated terminal structures is minimal (see Fig. 2.4). The lower and higher temperature peaks in the unmixed curves are typical of pure poly(vinyl chloride) and

Figure 2.21. Diagramatic representation of degradation of pairs of polymers in TVA using twin limbed tubes. In the 'unmixed' situation the polymers are degraded simultaneously but in separate limbs. In the 'mixed' situation each limb contains a blend of the two polymers. Weights and sample thicknesses are constant. (Reproduced by kind permission of *Developments in Polymer Degradation – 1*, ed. N. Grassie, App. Sci. Pub., London, 1977, p. 176.)

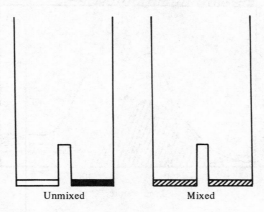

Unmixed   Mixed

poly(methyl methacrylate) respectively. The following changes have oc-
cured on mixing:

(*a*) The major products are the same but product analysis shows that
minor products originating from the poly(methyl methacrylate) are formed,
namely methanol and methyl chloride.

The area between the $-100\,^{\circ}\mathrm{C}$ and $-196\,^{\circ}\mathrm{C}$ curves, shaded in Fig. 2.22,
results from hydrogen chloride. The area between $-45\,^{\circ}\mathrm{C}$ and $-75\,^{\circ}\mathrm{C}$ is
due to methyl methacrylate but there is separation of $-75\,^{\circ}\mathrm{C}$ and $-100\,^{\circ}\mathrm{C}$
curves clearly due to new products.

(*b*) Methyl methacrylate is produced nearly $100\,^{\circ}\mathrm{C}$ lower concurrently
with hydrogen chloride.

(*c*) Methyl methacrylate occurs in two stages, the first at a lower
temperature as in (*b*), the second at a higher temperature than the unmixed

Figure 2.22. TVA behaviour of poly(methyl methacrylate) and poly(vinyl chloride)
mixed (*a*) and unmixed (*b*). Polymer ratio 50/50 by weight. Trap temperatures: ——
$0\,^{\circ}\mathrm{C}$ and $-45\,^{\circ}\mathrm{C}$; ····· $-75\,^{\circ}\mathrm{C}$; – – – $-100\,^{\circ}\mathrm{C}$; –·–·– $-196\,^{\circ}\mathrm{C}$. (Reproduced by
kind permission of *Developments in Polymer Degradation – 1*, ed. N. Grassie, App.
Sci. Pub., London, 1977, p. 181.)

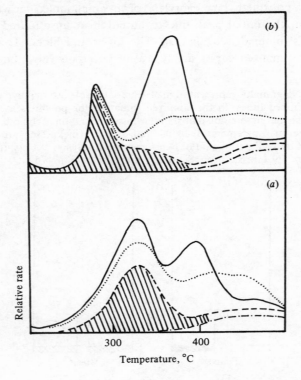

polymers. Thus there are both destabilising and stabilising processes affecting the poly(methyl methacrylate) when poly(vinyl chloride) is present.

These results can only be explained in terms of two interactive reactions. The first involves attack by chlorine atoms, which are intermediates in the dehydrochlorination of poly(vinyl chloride), on poly(methyl methacrylate), reaction (2.50).

$$
\begin{array}{c}
\text{CH}_3 \qquad \text{CH}_3 \qquad \text{CH}_3 \\
\sim\!\!\text{C}-\text{CH}_2-\text{C}-\text{CH}_2-\text{C}-\text{CH}_2\!\!\sim \quad \xrightarrow{\text{Cl·}} \quad \sim\!\!\text{C}-\text{CH}_2-\text{C}-\dot{\text{C}}\text{H}-\text{C}-\text{CH}_2\!\!\sim \\
\text{COOCH}_3 \ \text{COOCH}_3 \ \text{COOCH}_3 \qquad\qquad \text{COOCH}_3 \ \text{COOCH}_3 \ \text{COOCH}_3
\end{array}
$$

$$
\text{Monomer} \longleftarrow \sim\!\!\overset{\text{CH}_3}{\underset{\text{COOCH}_3}{\text{C}}}-\text{CH}_2 \ + \ \overset{\text{CH}_3}{\underset{\text{COOCH}_3}{\text{C}}}=\text{CH}-\overset{\text{CH}_3}{\underset{\text{COOCH}_3}{\text{C}}}-\text{CH}_2\!\!\sim
$$

(2.50)

The second process, reaction (2.51), is the reaction of hydrogen chloride with the ester groups of the poly(methyl methacrylate)

$$
\sim\!\!\text{CH}_2-\overset{\text{CH}_3}{\text{C}}-\text{CH}_2-\overset{\text{CH}_3}{\text{C}}\!\!\sim \longrightarrow \sim\!\!\text{CH}_2-\overset{\text{CH}_3}{\text{C}}-\text{CH}_2-\overset{\text{CH}_3}{\text{C}}\!\!\sim \ + \ \text{CH}_3\text{OH} + \text{CH}_3\text{Cl}
$$

(2.51)

In the first process, the highly-reactive chlorine atoms abstract hydrogen atoms from the polymer to give radicals which ultimately unzip at the poly(vinyl chloride) degradation temperature. Reaction with hydrogen chloride, on the other hand gives methanol and methyl chloride at poly(vinyl chloride) degradation temperatures but also anhydride groups which stabilise the molecules by blocking unzipping at higher temperatures.

Poly(vinyl chloride) – poly(vinyl acetate)

The mixed/unmixed TVA comparison for the poly(vinyl chloride)–poly(vinyl acetate) system is shown in Fig. 2.23. The shaded area, between the $-196\,°C$ and $-100\,°C$ traces is due to hydrogen chloride. On the other hand, the area between the $-100\,°C$ and $0\,°C$ curves is due to products of deacetylation of poly(vinyl acetate), mostly acetic acid.

For the unmixed situation, dehydrochlorination and deacetylation are well separated, dehydrochlorination occurring much more readily. In the blend, production of acetic acid occurs concurrently with hydrochlorination. But the dehydrochlorination rate maximum occurs earlier in the blend so that both polymers are destabilised.

This is rather similar to the behaviour of copolymers of vinyl chloride and vinyl acetate in which it was noted that small amounts of either monomer caused instability in the other polymer. That was accounted for in terms of catalysis of the decomposition of vinyl acetate units by adjacent vinyl chloride units. In the blend such interaction would have to take place between different polymer molecules across a phase boundary. Direct acid catalysis therefore seems more likely, reaction (2.52), hydrogen chloride

Figure 2.23. TVA behaviour of poly(vinyl chloride) and poly(vinyl acetate) mixed (*a*) and unmixed (*b*). Polymer ratio 50/50 by weight. Trap temperatures as in Fig. 2.22. (Reproduced by kind permission of *Developments in Polymer Degradation – 1*, ed. N. Grassie, App. Sci. Pub., London, 1977, p. 186.)

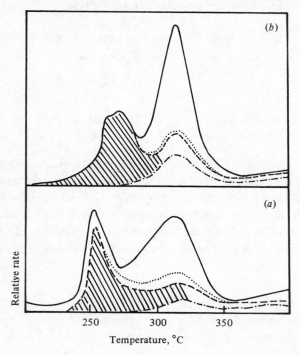

produced in the poly(vinyl chloride) phase diffusing into the poly(vinyl acetate) phase to catalyse loss of acetic acid.

$$\text{~CH}_2-\text{CH}-\text{CH}-\text{CH}\text{~} \quad\quad \text{~CH}_2-\text{CH}=\text{CH}-\text{CH}\text{~}$$

(2.52)

Acetic acid then diffuses into the poly(vinyl chloride) phase to catalyse loss of hydrogen chloride.

(2.53)

Once reaction is initiated in this way it will continue along the chain due to allylic activation.

In each of these two poly(vinyl chloride) blends the interactions which occur between the two polymers during degradation depend upon diffusion of small molecular species, radical or molecule, across the phase boundary. Experiments on a wider range of polymers have indicated that this is generally so.

Suggested further reading

1. N. Grassie, Thermal Degradation of Polymers, *Chemical Reactions of Polymers*, ed, E. M. Fettes, Interscience 1964, pp. 565–644.
2. N. Grassie, Degradation of Polymers, *Encyclopaedia of Polymer Science and Technology*, Interscience, 1966, pp. 647–716.
3. H. H. G. Jellinek (ed.), *Aspects of Degradation and Stabilisation of Polymers*, Elsevier, 1978.
4. R. T. Conley (ed.), *Thermal Stability of Polymers*, Dekker, 1970.
5. C. David, Thermal Degradation of Polymers, *Comprehensive Chemical Kinetics*, Vol. 14, ed. C. H. Bamford and C. F. H. Tipper, Elsevier, 1975.
6. N. Grassie, Degradation, *Polymer Science – A Materials Science Handbook*, Vol. 2, ed. A. D. Jenkins, North Holland, 1972, chapter 22.
7. W. Schnabel, *Polymer Degradation*, Hanser, 1981.
8. I. C. McNeil, Application of Thermal Volatilisation Analysis to Studies of Polymer Degradation, *Developments in Polymer Degradation – 1*, ed. N. Grassie, App. Sci. Pub., London, 1977.
9. N. Grassie, The Pyrolysis of Acrylonitrile Homopolymers and Copolymers, *Developments in Polymer Degradation – 1*, ed. N. Grassie, App. Sci. Pub., London, 1977.
10. I. C. McNeil, The Thermal Degradation of Polymer Blends, *Developments in Polymer Degradation – 1*, ed. N. Grassie, App. Sci. Pub., London, 1977.

3

Photo-degradation

3.1. Introduction

Sunlight was soon recognized as an important factor in the deteriorative ageing and weathering processes which occur in commercial polymers. The reasons for this are readily understood. The wavelength of the radiation from the sun which reaches the earth's surface extends from the infra-red (>700 nm), through the visible spectrum (approximately 400–700 nm) into the ultra-violet (<400 nm) with a cut-off at approximately 300 nm depending upon atmospheric conditions. The energies of 700, 400 and 300 nm photons are approximately 170, 300 and 390 kJ mol^{-1} respectively. The strengths of C–C and C–H bonds are approximately 420 and 340 kJ mol^{-1} respectively although they may be very much less in certain environments, for example in the neighbourhood of aromatic or unsaturated structures. Thus it is clear that the energy of the quanta of the UV and possibly of the visible components of sunlight is sufficient to break chemical bonds and that the shorter wavelengths will be the more effective.

Of course it is not enough that sufficiently energetic quanta are available. Chromophoric groups are necessary to absorb the incident radiation. In polymers, these are usually unsaturated structures such as carbonyl, ethylenic or aromatic groups. The absorption of energy and its transfer to the bond to be broken may be described as the photophysical aspect of photo-degradation. This is a very large subject in its own right and beyond the scope of the present treatment, which will be concerned only with the chemical processes which occur from the time at which the initial bond scission occurs.

Because absorption of radiation is an essential first step to photo-degradation, strongly absorbed radiation will be attenuated as it passes through the polymer and reaction will be concentrated in the surface layers. It is for this reason that a 'skin effect' is frequently observed in photo-initiated reactions.

The first chemical step in photo-degradation is usually homolytic bond scission to form free radicals. These radicals will normally react rapidly with any oxygen present. In this way, visible and especially UV radiation are particularly effective initiators of oxidation.

It will become clear that photo-degradation has close associations with thermal degradation (chapter 2) and degradation induced by high-energy radiation (chapter 7) as well as photo-oxidation. It is because of its central position in polymer degradation that the basic facts of photo-degradation are presented separately in this chapter and photo-oxidation is dealt with at greater length in chapter 4.

Much of the work described in this chapter has made use of 254 nm radiation. This is well beyond the range of sunlight but it is conveniently produced in high intensity from medium pressure mercury vapour lamps and causes reactions to proceed at conveniently fast rates for experimentation.

3.2. Polyolefins

Since unsaturated chromophoric groups are necessary for absorption, it is to be anticipated that fully saturated polymers like the polyolefins should be immune to photo-degradation. Yet most polymers are susceptible to photo-degradation to some extent. This must be due to small amounts of impurities or structural abnormalities in the macromolecular structure.

Both chain scission and cross linking have been reported to occur when polyethylene is irradiated. This can easily be accounted for if it can be assumed that radicals of type I are present in the irradiated polymer.

$$\sim\sim CH_2-CH-CH_2-CH_2\sim\sim \atop \sim\sim CH_2-CH-CH_2-CH_2\sim\sim \qquad (3.1)$$

$$\sim\sim CH_2-\overset{\cdot}{C}H-CH_2-CH_2\sim\sim$$
$$\text{I}$$

$$\sim\sim CH_2-CH=CH_2 + \cdot CH_2\sim\sim \qquad (3.2)$$
$$\text{II}$$

By the application of esr spectroscopy both radicals I and II have been shown to be present in irradiated polyethylene. Radical I may be formed by hydrogen transfer from a polyethylene molecule to any other radical present in the system.

The fact that thermally labile structural abnormalities exist in polyethylene has already been referred to in chapter 2. Of these, carbonyl groups are liable to be most photo-labile and experiments on copolymers of ethylene and carbon monoxide, III have confirmed this. On exposure to UV radiation, ketones of this type undergo Norrish Type I and Type II reactions.

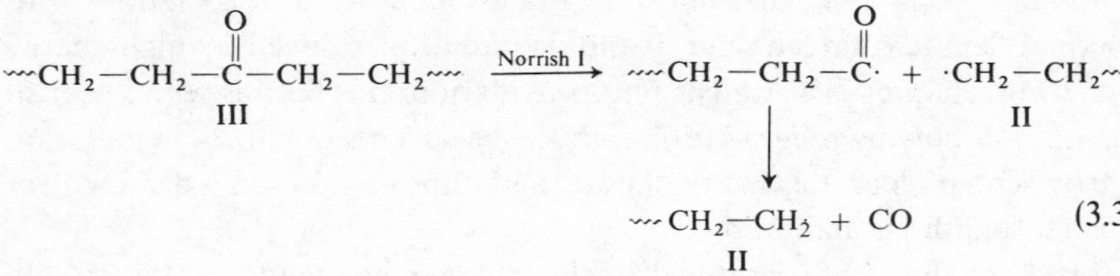

$$\begin{array}{c} O \\ \parallel \\ \sim CH_2-CH_2-C-CH_2-CH_2\sim \end{array} \xrightarrow{\text{Norrish I}} \begin{array}{c} O \\ \parallel \\ \sim CH_2-CH_2-C\cdot \end{array} + \cdot CH_2-CH_2\sim$$

III   II

$$\sim CH_2-CH_2 + CO \qquad\qquad (3.3)$$

II

$$\begin{array}{c} O\cdots H \\ \diagdown \quad \diagup \\ \sim C \qquad CH\sim \\ \diagup \quad \diagdown \\ H_2C-CH_2 \end{array} \xrightarrow{\text{Norrish II}} \begin{array}{c} OH \\ \mid \\ \sim C \\ \diagdown \\ CH_2 \end{array} + CH_2{=}CH\sim$$

III

$$\begin{array}{c} O \\ \parallel \\ \sim C-CH_3 \end{array} \qquad\qquad (3.4)$$

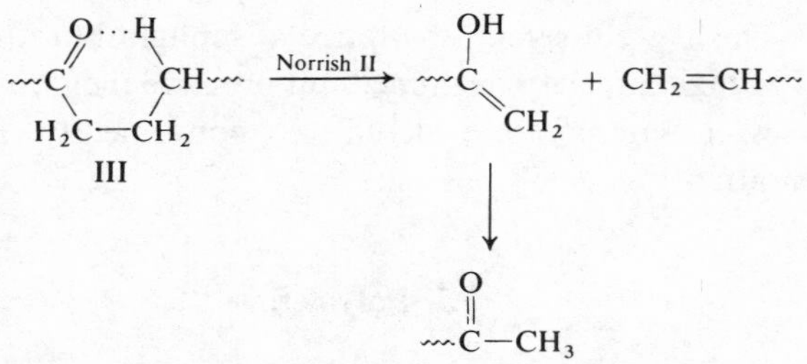

In the first, chain scission occurs adjacent to the carbonyl group to form radicals of type II directly. The second is a non-radical reaction which results in chain scission by way of a six-membered ring transition state yielding a methyl ketone and vinyl unsaturation.

Owing to the relatively high reactivity of the tertiary hydrogen atom in polypropylene, this polymer is even more susceptible than polyethylene to small amounts of oxidation during preparation, storage and processing. The carbonyl groups thus formed and their hydroperoxide precursors are both decomposed by visible/UV radiation forming radicals. This will be discussed in the following chapter. In the absence of oxygen these radicals will attack the polymer, especially the tertiary hydrogen atoms, forming in chain radicals, e.g. IV,

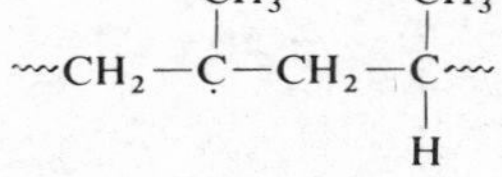

$$\begin{array}{c} CH_3 \qquad CH_3 \\ \mid \qquad\quad \mid \\ \sim CH_2-C-CH_2-C\sim \\ \qquad\qquad\quad \mid \\ \qquad\qquad\quad H \end{array}$$

IV

so that polypropylene, like polyethylene, undergoes both chain scission and cross linking in reactions analogous to (3.1) and (3.2) above.

3.3. Polyketones

Although the rapid chain scission which occurs in poly(methyl vinylketone), V, poly(methyl isopropenyl ketone), VI and poly(phenyl vinyl ketone), VII under UV radiation has been recognised

$$
\begin{array}{ccc}
\underset{\text{V}}{
\begin{array}{c}
\text{H} \\
\text{---CH}_2\text{---C---} \\
\text{C}{=}\text{O} \\
\text{CH}_3
\end{array}}
&
\underset{\text{VI}}{
\begin{array}{c}
\text{CH}_3 \\
\text{---CH}_2\text{---C---} \\
\text{C}{=}\text{O} \\
\text{CH}_3
\end{array}}
&
\underset{\text{VII}}{
\begin{array}{c}
\text{H} \\
\text{---CH}_2\text{---C---} \\
\text{C}{=}\text{O} \\
\bigcirc
\end{array}}
\end{array}
$$

for a considerable time it is only comparatively recently that they have attracted commercial interest, particularly in the context of the production of photo-degradable polymers. The principle is that when commercial polymers incorporating a small proportion of copolymerised ketone are discarded after use, their molecular weight should be reduced rapidly on exposure to sunlight. The friable, low molecular weight residue would then be reduced to a powder and scattered by the elements or destroyed by bacterial action to which the original high molecular weight material was immune.

In V and VI photolysis involves both Norrish I and Norrish II cleavage. In V the former is represented as follows:

$$
\underset{\text{V}}{
\begin{array}{c}
\text{---CH}_2\text{---CH---} \\
\text{C}{=}\text{O} \\
\text{CH}_3
\end{array}}
\xrightarrow{h\nu}
\begin{cases}
\text{---CH}_2\text{---}\overset{\cdot}{\text{C}}\text{H---} + \overset{\text{O}}{\underset{\|}{\cdot\text{C}}}\text{---CH}_3 \longrightarrow \text{CO} + \cdot\text{CH}_3 \\[2ex]
\cdot\text{CH}_3 + \text{---CH}_2\text{---CH---} \longrightarrow \text{---CH}_2\text{---}\overset{\cdot}{\text{C}}\text{H---} + \text{CO} \\
\phantom{\cdot\text{CH}_3 + \text{---CH}_2\text{---}}\text{C}{=}\text{O}
\end{cases}
$$

$$(3.5)$$

The methyl and acetyl radicals may abstract hydrogen atoms from the polymer to form methane and acetaldehyde which are, with carbon monoxide, the only significant volatile products from poly(methyl vinyl ketone).

Monomer is an additional product from poly(methyl isopropenyl ketone), especially at higher temperatures, and must be due to the ability of the chain terminal radicals to depropagate.

In poly(phenyl vinyl ketone), VII, the aromatic group has the effect of making Norrish II cleavage predominant, reaction (3.6).

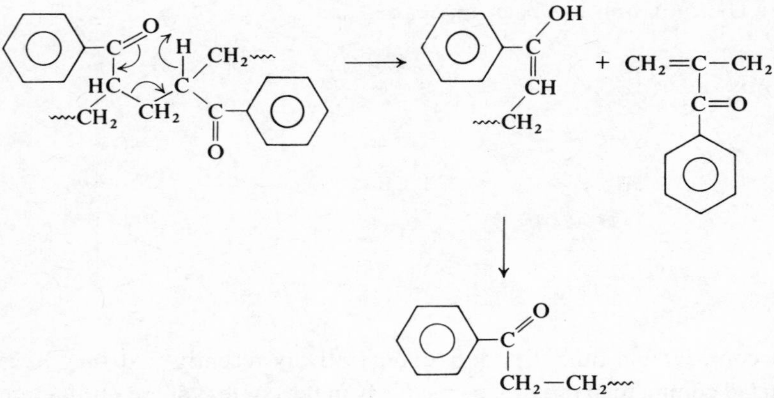

$$(3.6)$$

A small proportion of this monomer copolymerised into polystyrene acts as a potent photo-destabiliser for the latter.

3.4. Acrylates and methacrylates

It is interesting to consider why two classes of polymers as similar as the polyacrylates, VIII, and polymethacrylates, IX, should apparently behave so differently on UV irradiation.

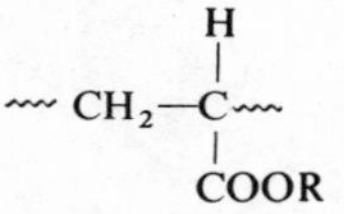

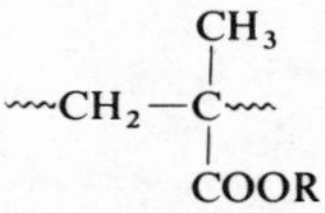

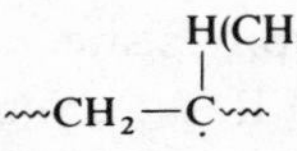

At ambient temperatures, poly(methyl methacrylate) undergoes chain scission, the molecular weight decreasing rapidly, while poly(methyl acrylate) becomes insoluble due to cross-linking of the polymer molecules. Nevertheless these reactions also have a great deal in common since the gaseous products of reaction, although they are produced in relatively small quantities, are identical. These are principally methyl formate, methanol, methane, hydrogen, carbon monoxide and carbon dioxide formed by decomposition of the ester side group. Thus it seems clear that the primary effect of UV radiation is to cause scission of the ester group giving the radical X and that the subsequent reactions of this radical determine the more obvious changes which occur. Thus polymethacrylate radicals predominantly undergo chain scission, reaction (3.7),

$$\text{~CH}_2-\underset{\underset{\text{COOCH}_3}{|}}{\overset{\overset{\text{CH}_3}{|}}{\text{C}}}-\text{CH}_2-\overset{\overset{\text{CH}_3}{|}}{\text{C}}\text{~} \longrightarrow \text{~CH}_2-\overset{\overset{\text{CH}_3}{|}}{\text{C}}=\text{CH}_2 + \cdot\underset{\underset{\text{COOCH}_3}{|}}{\overset{\overset{\text{CH}_3}{|}}{\text{C}}}\text{~}$$

(3.7)

while acrylate radicals predominantly combine, reaction (3.8).

$$\begin{array}{c}\text{~CH}_2-\overset{\overset{\text{COOCH}_3}{|}}{\text{CH}}-\text{CH}_2-\text{CH~}\\ \text{~CH}_2-\dot{\text{C}}\text{H}-\text{CH}_2-\underset{\underset{\text{COOCH}_3}{|}}{\text{CH}}\text{~}\end{array} \longrightarrow \begin{array}{c}\text{~CH}_2-\overset{\overset{\text{COOCH}_3}{|}}{\text{CH}}-\text{CH}_2-\text{CH~}\\ \text{~CH}_2-\text{CH}-\text{CH}_2-\underset{\underset{\text{COOCH}_3}{|}}{\text{CH}}\text{~}\end{array}$$ (3.8)

When thin films of copolymers of these two monomers, covering the whole composition range, are exposed to 254 nm radiation, only those copolymers containing more than 50 mole % of methyl methacrylate remain soluble. Rates of chain scission may be calculated from the change in molecular weight using equation (*i*) (see chapter 2)

$$\alpha = 1/P_t - 1/P_0 \qquad (i)$$

in which α is the number of scissions per monomer unit and P_0 and P_t are the number average chain lengths at zero time and time t. Copolymers containing less than 50 mole % of methyl methacrylate become progressively insoluble during irradiation. Sol–gel analyses have been carried out on these and the data analysed using the Charlesby–Pinner equation, (*ii*)

$$S + S^{1/2} = p_0/q_0 + 1/q_0 ut \qquad (ii)$$

in which S is the sol fraction, p_0 and q_0 are rates of scission and cross-linking, u is the number average degree of polymerisation of the starting material and t is the time of irradiation. Data, illustrated in Fig. 3.1, reveal a number of interesting facts. First, although the rate of cross-linking decreases with decreasing acrylate content of the copolymer as expected, an extrapolation suggests that it is effectively zero even in copolymers containing as much as 45 mole % of methyl acrylate units. Secondly, the shape of the chain scission curve is perhaps unexpected.

One might reasonably expect the rate of chain scission to decrease with decreasing methyl methacrylate content at high methacrylate contents but the minimum in the vicinity of 40 to 50 mole % methyl methacrylate and the subsequent rapid increase so that the rate of scission in pure poly(methyl acrylate) is of the same order as that in pure poly(methyl methacrylate) is surprising.

In solution, UV radiation does not cause insolubility in poly(methyl acrylate) or methyl acrylate rich copolymers as it does in film. It may be assumed that this is because cross-linking is inhibited in solution by the separation of the polymer molecules by molecules of the solvent. Equation (*i*) has been applied to molecular weight measurements made during irradiation of copolymers in solution in methyl acetate with the results shown in Fig. 3.2. The close similarity between the principal characteristics of the photo-degradation of methyl methacrylate/methyl acrylate copolymers in the solid phase and in solution is demonstrated by comparison of Figs. 3.1 and 3.2. The intensity of irradiation in each case was comparable.

A closer comparison reveals that while the rates of chain scission in poly(methyl acrylate) and methyl acrylate-rich copolymers in film and in solution are comparable, the rate of chain scission in methyl methacrylate-rich copolymers is very much greater in solution.

One obvious difference between the methyl acrylate and methyl methac-

Figure 3.1. Dependence of rates of scission and cross-linking on the composition of copolymers of methyl methacrylate and methyl acrylate irradiated as films. ● chain scission (equation (*i*)); ▲ chain scission (equation (*ii*)); ■ cross-linking (equation (*ii*)).

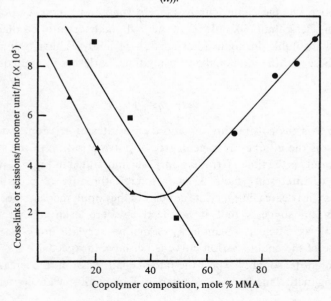

rylate ends of the copolymer composition range is that the glass transition temperatures of high methyl acrylate copolymers are below ambient temperature at which these film and solution photolyses were carried out, while the glass transition temperatures of high methacrylate copolymers lie well above ambient temperatures. The values of T_g for poly(methyl methacrylate) and poly(methyl acrylate) are 105 °C and 6 °C respectively. It might thus reasonably be predicted that the immobility of the polymer molecules in high methacrylate copolymer films would result in a high proportion of 'cage' recombination of the primary radical products. In solution, on the other hand, these primary radicals would more readily diffuse apart thus explaining the higher rate of degradation in solution. Because, at ordinary temperatures, high acrylate copolymers are above their glass transition temperatures there should be much less tendency to 'cage' recombination of radicals and rates of degradation in film should increase towards the rates in solution.

This theory was tested by measuring rates of photolytic scission of a high methyl methacrylate copolymer over a range of temperatures. Figure 3.3 illustrates the rapid increase in the rate of photolytic scission in the vicinity of the glass transition temperature which was 82 °C for this material.

But there are also fundamental differences in the nature of the volatile degradation products when methacrylate polymers are irradiated below and above the glass transition temperature. At 160 °C, which is above T_g but

Figure 3.2. Dependence of rates of scission on the composition of copolymers of methyl methacrylate and methyl acrylate irradiated in solution in methyl acetate.

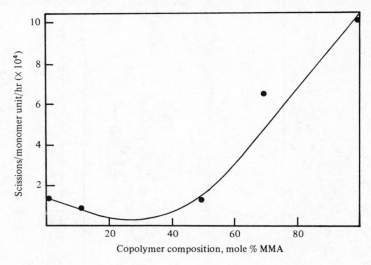

below the threshold for thermal degradation (approximately 200 °C) radicals are produced and unzipping of the chain end radicals occurs so that quantitative yields of monomer are obtained. Unlike the thermal reaction (chapter 2) this photo-initiated unzipping process is common to all methacrylate esters above T_g, even poly(t-butyl methacrylate) whose thermal degradation is almost exclusively ester decomposition.

This difference between the products of photo-degradation in film and solution on the one hand and in molten polymer on the other can be explained in terms of the equilibrium which exists between propagation and depropagation,

$$M_n^{\bullet} \rightleftharpoons M_{n-1}^{\bullet} + M \tag{3.9}$$

in which $M_n^{\bullet}$ and $M_{n-1}^{\bullet}$ are polymer radicals, n and $n-1$ monomer units in length, and M is a monomer molecule. At higher temperatures in molten polymer, conditions are such that monomer is continuously removed so that the equilibrium tends to the right and high yields of monomer are obtained. In film at ambient temperature, monomer cannot freely escape so that each radical will find itself effectively surrounded by a high concentration of monomer so that the monomer-producing reaction is inhibited. In solution the equilibrium concentration of monomer, which is

Figure 3.3. Dependence on temperature of the rate of chain scission under UV radiation of a 70/30, methyl methacrylate/methyl acrylate copolymer.

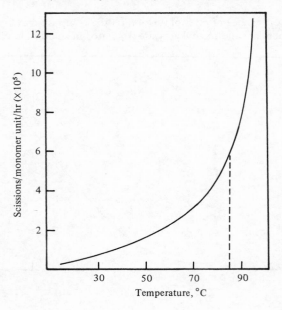

quite low, will be rapidly built up because solution experiments are invariably carried out in a closed system because of the volatility of the solvent.

These photolyses reactions of polymethacrylates, carried out above T_g and which result in high yields of monomer, have been described as photo-thermal reactions and are illustrated very well using the TVA technique (see chapter 2). A TVA curve for poly(methyl methacrylate) without irradiation is shown in Fig. 2.4. An identical experiment, but with the polymer irradiated, is illustrated in Fig. 3.4 which demonstrates that monomer production starts at a very much lower temperature. This demonstrates that evolution of monomer from PMMA is principally associated with producing radicals under appropriate conditions. These conditions must be such that monomer can escape readily as in molten polymer and that it is removed efficiently from the system so that the polymerisation/depolymerisation equilibrium is pushed far to the right.

3.5. Copolymers of methyl methacrylate and methyl vinyl ketone

The behaviour of copolymers of methyl methacrylate and methyl vinyl ketone, illustrated in Fig. 3.5, demonstrates how two photo-degradable monomers in a polymer chain can interact upon one another to produce a copolymer which is even more readily photo-degradable than either of the homopolymers.

It has been pointed out that in poly(methyl vinyl ketone) there is considerable scope for energy transfer from the carbonyl group initially

Figure 3.4. TVA curves for poly(methyl methacrylate) simultaneously irradiated with 253.7 nm radiation. Trap temperatures: —— 0 °C; ----- −45 °C; − − − −75 °C; ····· −100 °C. (Reproduced by kind permission of *J. Polym. Sci. Chem. Ed.*, **15**, 251, 1977.)

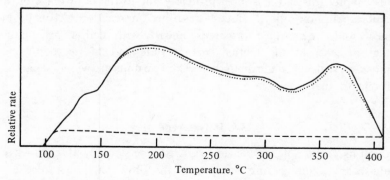

Polymer degradation and stabilisation

excited by the radiation to adjacent groups in the polymer molecules. This provides a mechanism for conversion of electronic to vibrational energy. This is probably one of the principal reasons why the quantum yield for chain scission is considerably less than unity. When MMA groups are introduced into the PMVK, this energy transfer will be inhibited, thus increasing the rate of chain scission. It has also been suggested that seven membered ring interactions involving the MMA group, reaction (3.10), can lead to efficient chain scission.

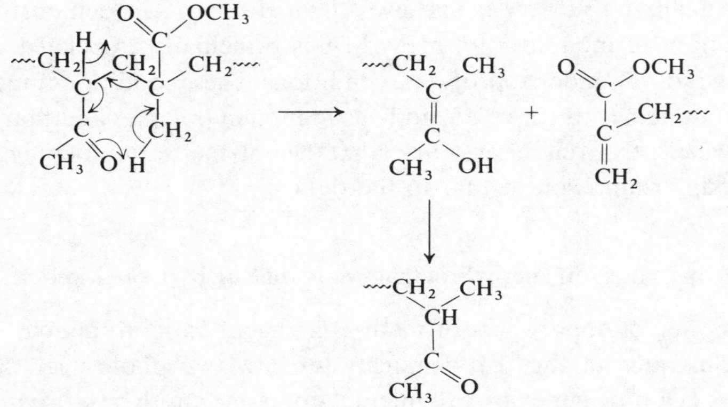

$$(3.10)$$

The data in Fig. 3.5, which cover the entire range of copolymer composition, confirms this accelerating effect by MMA units and show that it continues up to approximately 80% of MMA groups in the copolymer.

Figure 3.5 also demonstrates that, at the other end of the composition scale, small amounts of methyl vinyl ketone increase the rate of scission of poly(methyl methacrylate) dramatically. It seems unlikely that this is to be associated with an increase in the rate of the radical processes which occur in poly(methyl methacrylate) and which are described in a previous section of this chapter. Instead it is most probably due to the occurrence of the intramolecular non-radical process described in reaction (3.10). Its incidence should be expected to increase linearly with methyl vinyl ketone content at least at the methyl methacrylate end of the copolymer composition scale when the methyl vinyl ketone units are widely separated from one another.

3.6. Polystyrene

Through its chromophoric benzene ring, styrene strongly absorbs wavelengths below about 290 nm. In spite of this, however, polystyrene is a comparatively radiation stable polymer because the benzene ring also acts as an 'energy sink' through which the energy of the radiation can be dissipated. The strength of the absorption, especially of the lower, more energetic and thus more photolytically active wavelengths, leads to marked attenuation of the radiation as it passes through the polymer so that photolysis of polystyrene tends to be concentrated at the surface.

Photolysis manifests itself in three ways: by evolution of hydrogen, development of insolubility and discolouration. ESR spectra indicate that radical XI is present in the irradiated polymer and this is not surprising since the tertiary carbon–hydrogen bonds should be much weaker than

$$\text{--- CH}_2\text{--}\overset{|}{\text{C}}\text{--CH}_2\text{---}$$

XI

Figure 3.5. Effect of copolymer composition on the rate of chain scission of copolymers of methyl methacrylate and methyl vinyl ketone during photodegradation in solution in methyl acetate. (Reproduced by kind permission of *Polym. Degrad. and Stab.*, **3**, 45, 1980.)

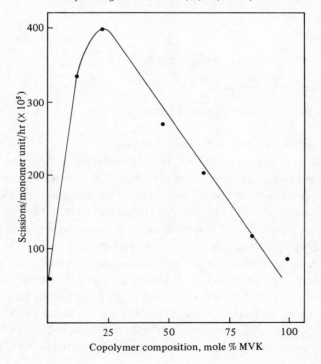

either the methylene or aromatic bonds. Hydrogen molecules would be formed either by combination of hydrogen atoms in pairs or by abstraction of hydrogen atoms from the polymer structure. The development of insolubility would be accounted for by combination of pairs of XI radicals. Abstraction of a hydrogen atom from the carbon atom adjacent to the radical centre would be energetically favourable and would result in

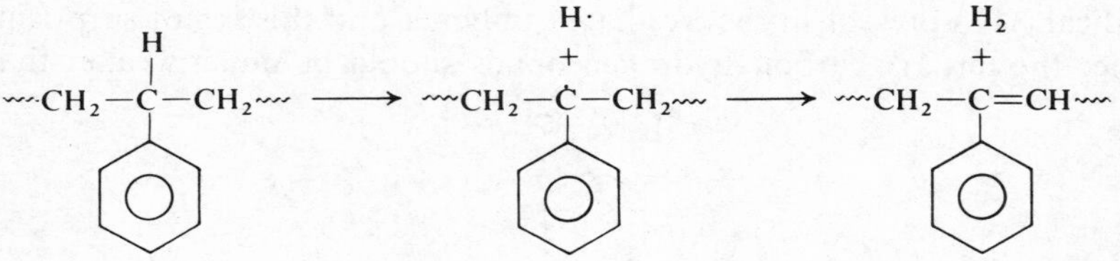

$$\text{(3.11)}$$

unsaturation in conjugation with the benzene ring. The existence of conjugated structures of this kind in photolysed polystyrene is supported by spectral evidence. The mechanism may thus be summarised as follows,

$$RH \xrightarrow{h\nu} R^{\bullet} + H^{\bullet} \tag{3.12}$$

$$H + RH \longrightarrow H_2 + R^{\bullet} \tag{3.13}$$

$$H^{\bullet} + R^{\bullet} \longrightarrow H_2 + \text{unsaturation} \tag{3.14}$$

$$H^{\bullet} + H^{\bullet} \longrightarrow H_2 \tag{3.15}$$

$$R^{\bullet} + R^{\bullet} \longrightarrow R-R \text{ (cross-linking)} \tag{3.16}$$

in which RH represents a polymer molecule.

The relative rates of reaction (3.13), (3.14) and (3.15) will depend upon the mobility of hydrogen atoms. Thus if the escape of hydrogen atoms into the gas phase is inhibited by the presence of nitrogen then reaction (3.15) will be inhibited and reactions (3.13) and (3.14) thus accelerated. Figure 3.6 shows that the development of unsaturation, as measured by absorption at 240 nm does indeed increase with nitrogen pressure.

Industrially, the most important adverse manifestation of ageing of polystyrene is the development of yellow colouration. Although this was traditionally associated with oxidation it has been shown to occur even under vacuum. It has been suggested that unsaturation in the polymer chain, produced as described above, will render the α-hydrogen atom in XII, which is also a tertiary hydrogen atom, even more labile so that it will

$$\sim\!\!C\!\!=\!\!CH\!-\!\underset{\underset{\displaystyle\bigcirc}{|}}{\overset{\displaystyle H}{\underset{|}{C}}}\!-\!CH_2\!\sim$$

XII

react more readily than a similar bond in an unchanged region of the polymer molecule. Thus reaction of sequences of adjacent units would be favoured and conjugation result.

The main features of the photolysis of poly(α-methyl styrene) and poly(p-methyl styrene) are consistent with this overall mechanism for the photolysis of polystyrene. Thus poly(p-methyl styrene) which retains the tertiary hydrogen atoms, like polystyrene becomes rapidly insoluble, liberates hydrogen and discolours, while poly(α-methyl styrene), in which the tertiary hydrogen atoms have been replaced by methyl groups, does not discolour or cross link but undergoes rapid chain scission and liberates monomer at higher temperatures.

Figure 3.6. Effect of pressure of nitrogen on the rate of development of unsaturation during irradiation of polystyrene with 253.7 nm radiation. (Reproduced by kind permission of *Polymer Science*, ed. A. D. Jenkins, North Holland Publishing Company, 1972, p. 1522.)

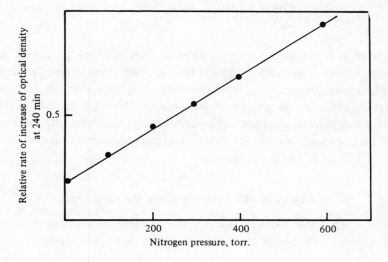

3.7. Other vinyl polymers

Polyacrylonitrile and poly(vinyl chloride) should not be expected to absorb, at least in the near UV. The fact that they do, and that this is followed by degradation, is a clear indication that chromophoric impurities are present which may be carbonyl groups due to oxidation or unsaturated structures built into the polymeric structure during the polymerisation process. Poly(vinyl acetate) and poly(vinyl alcohol), which is made by hydrolysis of poly(vinyl acetate) and always incorporates residual acetate groups, absorb through the acetate carbonyl, although poly(vinyl alcohol) is also known to incorporate additional carbonyl groups in the main chain.

Like the polyacrylates, with which they have tertiary carbon–hydrogen bonds in common, these four polymers are all subject to both cross-linking and chain scission under UV radiation in absence of oxygen. Cross-linking predominates in the solid phase so that they all become insoluble, but it has been reported that for polyacrylonitrile in solution chain scission predominates and the molecular weight is rapidly reduced.

The volatile products from polyacrylonitrile XIII, are hydrogen and hydrogen cyanide, suggesting that both of the bonds indicated

$$
\begin{array}{c}
\text{H} \\
\text{--}\!\!+\!\!\text{--} \\
\sim\!\!\text{CH}_2\!-\!\text{C}\!\sim \\
\text{--}\!\!+\!\!\text{--} \\
\text{CN}
\end{array}
$$

XIII

are broken. Unsaturation appears to be mostly due to terminal structures formed by chain scission, reaction (3.17),

$$
\sim\!\!\text{CH}_2\!-\!\overset{\cdot}{\text{C}}\!-\!\text{CH}_2\!-\!\underset{\underset{\text{CN}}{|}}{\text{CH}}\!\sim \longrightarrow \sim\!\!\text{CH}_2\!-\!\underset{\underset{\text{CN}}{|}}{\text{C}}\!=\!\text{CH}_2 + \underset{\underset{\text{CN}}{|}}{\cdot\text{CH}}\!\sim \quad (3.17)
$$

rather than conjugated sequences as in polystyrene and poly(vinyl chloride). Traces of monomer suggests that to some small extent the chain terminal radical is capable of depolymerising. Poly(vinyl chloride) liberates hydrogen chloride and conjugated unsaturation appears, behaviour closely reminiscent of its thermal degradation. Poly(vinyl acetate) produces acetic acid, carbon monoxide, carbon dioxide and methane suggesting scission of acetate groups as the initial step.

3.8. Polymers with heteroatoms in the main chain

The photolysis reactions of a number of important commercial polymers with hetero-atoms in the main chain have also been investigated.

Polyoxymethylene undergoes chain scission, the radicals depolymerising to some extent to form the monomer, formaldehyde.

$$\sim\sim CH_2-O-CH_2-O-CH_2-O\sim\sim \longrightarrow \sim\sim CH_2-O-CH_2-O\cdot + \cdot CH_2-O$$

$$\downarrow$$

$$\sim\sim CH_2-O\cdot + CH_2O.$$

(3.18)

The products of photolysis of formaldehyde, carbon monoxide and hydrogen, are also present in significant quantities among the volatile products.

Polycarbonates are relatively stable to photodegradation, probably, as in polystyrene, due to the 'energy sink' function of the aromatic rings. Random chain scission occurs at a very slow rate with liberation of carbon dioxide:

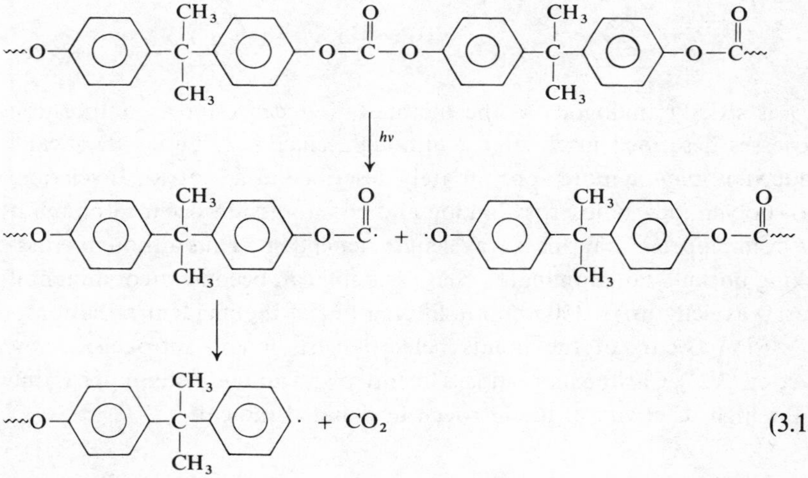

(3.19)

The radicals formed in this sequence can undergo a series of isomerisation and recombination processes so that new structures become incorporated into the polymer molecules.

3.9. Condensation polymers

Chain scission is also predominant in poly(ethylene terephthalate), XIV. The initial break can apparently occur in the ester group at any one of the three points indicated by the dotted lines. Carbon dioxide and carbon

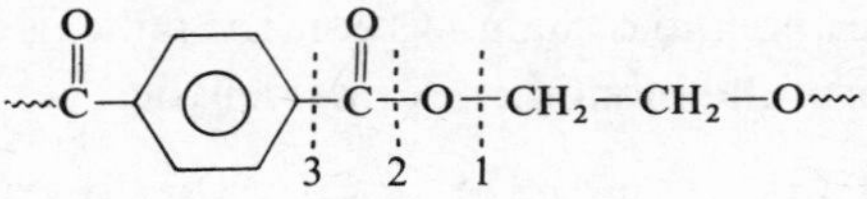

XIV

monoxide may be liberated following scission at points 1 and 2 and all of the radicals formed may abstract hydrogen from elsewhere in the system. An alternative chain scission process, reaction 3.20, is indicated by the fact that vinyl groups are formed in the degrading polymer.

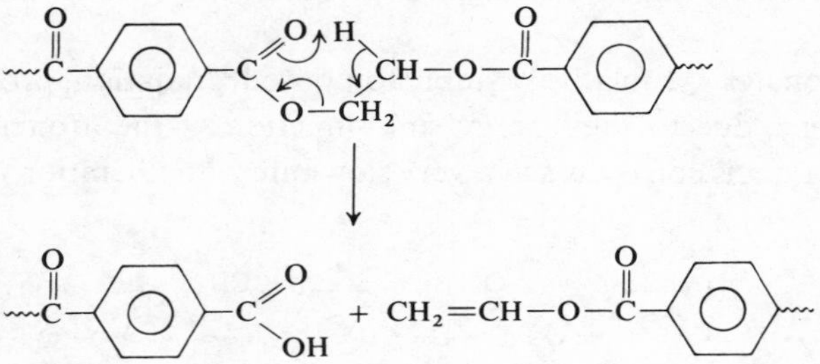

(3.20)

This is strictly analogous to the thermally induced ester decomposition processes described in chapter 2 although, since it is photo-chemically induced, it may be more appropriately described as a Norrish II scission.

In polyamides, both cross-linking and chain scission occur although it has been reported that this is wavelength dependent. Thus although cross-linking normally predominates, chain scission can become predominant if short wavelengths (< 300 nm) are filtered out of the incident radiation.

Scission occurs at the bonds between carbon and nitrogen atoms, reaction (3.21). Carbon monoxide is liberated and amine groups appear due to the high reactivity of the nitrogen terminated radical.

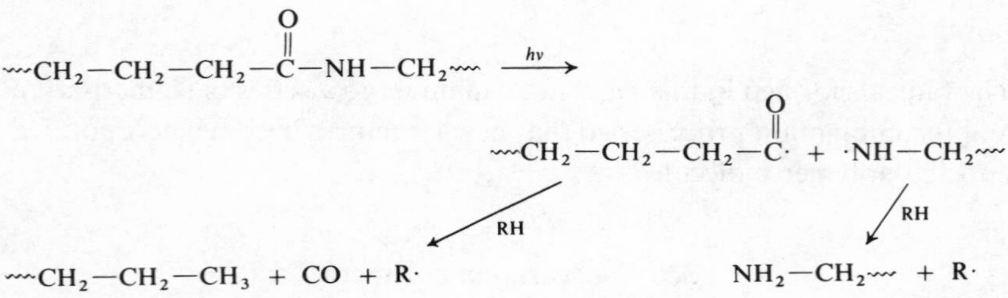

(3.21)

Crosslinking follows abstraction of hydrogen atoms from the methylene groups, the most reactive of which are those adjacent to the NH groups.

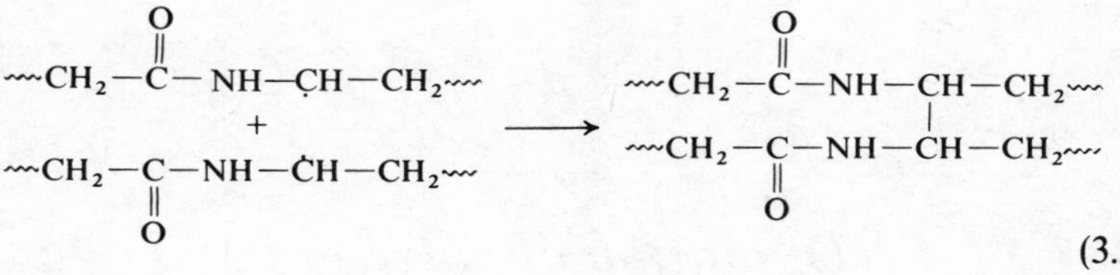

$$\begin{array}{c}
\overset{\text{O}}{\underset{\parallel}{\text{~~CH}_2-\text{C}-\text{NH}-\text{CH}-\text{CH}_2\text{~~}}} \\
+ \\
\overset{}{\underset{\overset{\parallel}{\text{O}}}{\text{~~CH}_2-\text{C}-\text{NH}-\text{CH}-\text{CH}_2\text{~~}}}
\end{array}
\longrightarrow
\begin{array}{c}
\overset{\text{O}}{\underset{\parallel}{\text{~~CH}_2-\text{C}-\text{NH}-\text{CH}-\text{CH}_2\text{~~}}} \\
\overset{}{\underset{\overset{\parallel}{\text{O}}}{\text{~~CH}_2-\text{C}-\text{NH}-\text{CH}-\text{CH}_2\text{~~}}}
\end{array}$$

$$(3.22)$$

Suggested further reading

1. G. Geuskens, Photo-degradation of Polymers, *Comprehensive Chemical Kinetics*, Vol. 14, ed. C. H. Bamford and C. F. H. Tipper, Elsevier, 1975.
2. W. Schnabel and J. Kiwi, Photo-degradation, *Aspects of Degradation and Stabilisation of Polymers*, ed. H. H. G. Jellinik, Elsevier, 1978.
3. R. B. Fox, Photo-degradation of High Polymers, *Progress in Polymer Science*, Vol. 1, ed. A. D. Jenkins, Pergamon, 1967.
4. B. Ranby and J. F. Rabek, *Photo-degradation, Photo-oxidation and Photo-stabilisation of Polymers*, Wiley, 1975.

4

Oxidation of polymers

4.1 Autoxidation

Although all polymers degrade at high temperatures in the absence of air, degradation is almost always faster in the presence of oxygen. Oxidation of hydrocarbons is normally auto-accelerating, i.e. the rate is slow or even negligible at first but gradually accelerates, often to a constant value (Fig. 4.1). The addition of an initiator normally removes the slow auto-accelerating induction time and antioxidants and stabilisers extend it.

4.1.1. The oxidation chain reaction

Ground state oxygen is unusual in that it exists in the triplet state, i.e. it is a diradical (I). Although excited singlet oxygen (II)

$$\dot{O}-\dot{O} \qquad\qquad O{=}O$$
$$\text{I} \qquad\qquad\quad \text{II}$$

can be important as an autoxidation initiator under certain circumstances, oxygen normally reacts with organic compounds in a radical chain reaction involving the ground state. Each cyclical sequence of reactions (4.1) and (4.2) absorbs one molecule of oxygen and leads to the formation of a hydroperoxide.

$$R\cdot + O_2 \rightarrow ROO\cdot \qquad\qquad (4.1)$$

$$ROO\cdot + RH \rightarrow ROOH + R\cdot. \qquad\qquad (4.2)$$

Since reaction (4.1) is a radical pairing process it has a low activation energy and occurs with high frequency. The second step (4.2) on the other hand involves the breaking of a carbon-hydrogen bond and has a higher activation energy. In most polymers at normal oxygen pressures, the rate of this step in the chain reaction determines the overall rate of oxidation. Exceptions to this general rule occur when the radical R· is strongly resonance-stabilised. For example, 2,6-dimethylhepta-2,5-diene(III) is oxidised to the radical IV, in which the unpaired electron is delocalised

Scheme 4.1 Formation of a resonance-stabilised radical during autoxidation.

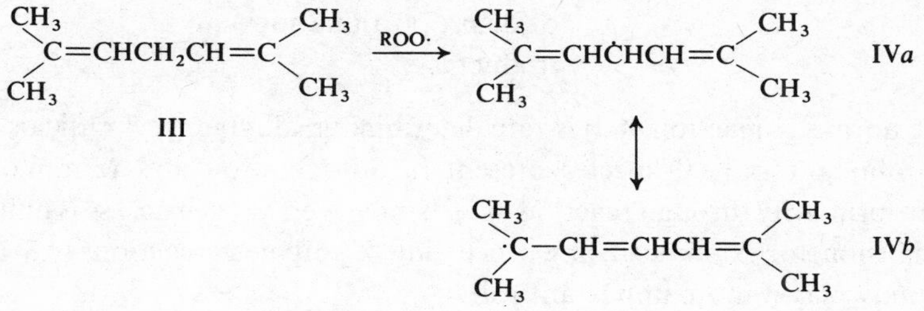

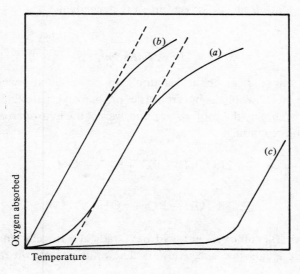

over five carbon atoms and to a lesser extent over four methyl groups by hyperconjugation. In this case reaction (4.2) is able to compete with reaction (4.1) at normal oxygen pressures.

4.1.2. Termination

The structure of the autoxidising hydrocarbon and the oxygen concentration determine which of the following termination steps lead to the removal of radicals from an autoxidising substrate:

Figure 4.1. Idealised oxygen absorption curves for oxidising polymers: (*a*) pure polymer; (*b*) polymer with added hydroperoxide; (*c*) polymer with added anti-oxidant.

$$2R\cdot \rightarrow R - R \tag{4.3}$$

$$R\cdot + ROO\cdot \rightarrow ROOR \tag{4.4}$$

$$2ROO\cdot \rightarrow ROOR + O_2 \tag{4.5}$$

(or disproportionation products).

Since normally reaction (4.2) is rate determining, alkylperoxyl radicals are the dominant radical species present in autoxidation and termination occurs primarily through reaction (4.5). If, however, oxygen access is limited by diffusion, for example during processing of polymers, reactions (4.3) and (4.4) may play a more important role.

The kinetic consequences of the relative efficiencies of reactions (4.1) and (4.2) is that for most hydrocarbons at ambient pressures, the rate of oxidation is given by equation (*i*), i.e. the oxidation is

$$\frac{-d[O_2]}{dt} = k_{4.2}k_{4.5}^{-1/2}r_i^{1/2}[RH] \tag{i}$$

where r_i is the rate of initiation.

independent of oxygen pressure. However, for highly oxidisable hydrocarbons or at low oxygen pressures, equation (*ii*) holds,

$$\frac{-d[O_2]}{dt} = k_{4.1}k_{4.4}^{-1/2}r_i^{1/2}[O_2] \tag{ii}$$

and the oxidation rate is oxygen pressure dependent.

4.1.3. Initiation

Although the radical chain reaction can be initiated by any radical generator (e.g. azobis-isobutyronitrile or benzoyl peroxide) initiation normally occurs by thermolysis or photolysis of the hydroperoxide formed in the chain reaction.

$$2ROOH \overset{\Delta H}{\rightarrow} RO\cdot + H_2O + \cdot OH \tag{4.6}$$

$$ROOH \overset{h\nu}{\rightarrow} RO\cdot + \cdot OH. \tag{4.7}$$

Since both the alkoxyl and hydroxyl radicals are efficient hydrogen abstracting agents, they effectively inject new radicals into the radical chain

(4.1), (4.2) by reaction (4.8). Transition metal ions, particularly cobalt, iron,

$$
\begin{array}{ccc}
\text{RO} \cdot & & \text{ROH} \\
\text{or} & \overset{\text{RH}}{\longrightarrow} & \text{or} \qquad + \text{R} \cdot \qquad (4.8) \\
\text{HO} \cdot & & \text{H}_2\text{O}
\end{array}
$$

manganese and copper are important catalysts for hydroperoxide decomposition. A combination of reactions (4.9) and (4.10) gives the normal

$$\text{ROOH} + \text{M}^+ \rightarrow \text{RO} \cdot + \text{OH}^- + \text{M}^{2+} \qquad (4.9)$$

$$\text{ROOH} + \text{M}^{2+} \rightarrow \text{ROO} \cdot + \text{H}^+ + \text{M}^+ \qquad (4.10)$$

stoichiometry of thermal decomposition in the absence of metal ions, reaction (4.6); the effect of the metal ions is, therefore, to reduce the activation energy of the hydroperoxide decomposition.

Since little or no hydroperoxide is present at the beginning of autoxidation, the rate of oxidation increases in the characteristic auto-accelerating manner as the concentration of hydroperoxide builds up in the system (Fig. 4.1, curve (*a*)). Deliberate addition of hydroperoxides (curve (*b*)) or of transition metal ions reduces or removes the auto-accelerating part of the curve giving, for a short time at least, a linear rate of oxidation due to a stationary hydroperoxide concentration; this results from the fact that under these conditions the rate of breakdown of hydroperoxide by reactions (4.6) or (4.7) is equal to its rate of formation by reaction (4.2).

Auto-acceleration is generally observed during both the thermal and photo-oxidation of hydrocarbon polymers. However, commercial hydro-carbon polymers are not 'pure' in the chemical sense. Not only do they contain chemical modifications in the polymer backbone due to side reactions occurring during polymerisation, but they may also contain catalyst residues (e.g. transition metal ions), cross-links (rubbers) or oxygen-containing groups introduced by adventitious oxidation. These may play an important initiating role under both thermal and photo-oxidative conditions and departure from ideality will be discussed in more detail in later sections.

4.1.4. Chemical changes in polymers during oxidative degradation

It is rather fortunate that the main chemical changes which occur in polymers during degradation can be readily followed by normal spectroscopic techniques. In hydrocarbon polymers, by far the most useful in following the progress of oxidation is IR spectroscopy and the most useful groups to follow are in the region 1710–1735 cm^{-1}, shown typically for polypropylene in Fig. 4.2. This is generally a composite absorbance in which aldehydes (1735 cm^{-1}) and ketones (1720 cm^{-1}) are the main species present during the early stages of degradation but carboxylic acids (1710 cm^{-1}) predominate in the later stages. The presence of carboxylic acids is indicative of chain scission processes shown for polyethylene in scheme 4.2 and the extent to which these are present generally correlates well with physical changes in the polymer (see chapter 1). Alcohol groups are also present from an early stage but since these give a broad absorbance in the IR due to hydrogen bonding, they are less useful as a kinetic measure of degradation. In some polymers (notably polyethylene) hydroperoxides can be detected by IR (at 3550 cm^{-1}) at an early stage of degradation, but these species can also be chemically monitored by standard chemical procedures (e.g. iodimetry). Spectroscopic and chemical procedures have

Figure 4.2. Infra-red spectrogram of polypropylene during photo-oxidation in the hydroxyl (3420 cm^{-1}) and carbonyl (1720 cm^{-1}) regions. Numbers on the curves represent UV irradiation times (hours).

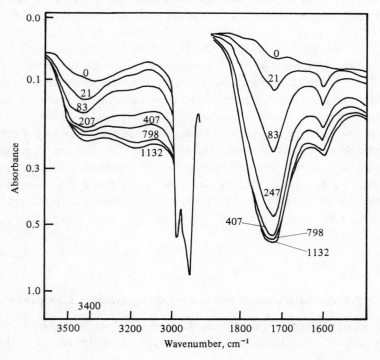

Scheme 4.2 Oxidative chain scission in polyethylene.

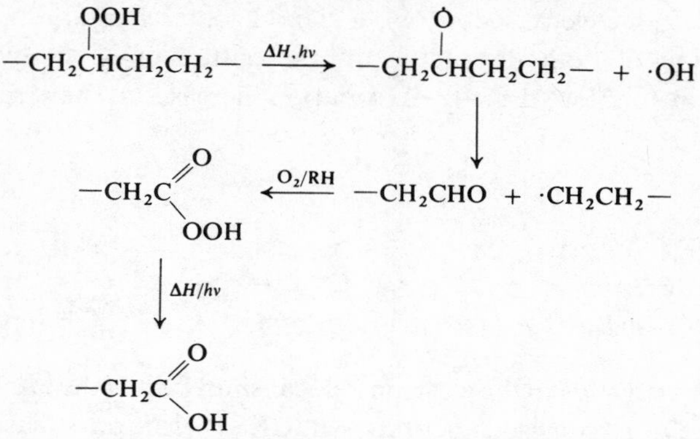

$$
\begin{array}{ccc}
\overset{\displaystyle OOH}{\underset{|}{}} & & \overset{\displaystyle O}{\underset{|}{}} \\
-CH_2\overset{|}{C}HCH_2CH_2- & \xrightarrow{\Delta H, h\nu} & -CH_2\overset{|}{C}HCH_2CH_2- \; + \; \cdot OH
\end{array}
$$

$$
\begin{array}{ccc}
& & \downarrow \\
-CH_2C\overset{\textstyle O}{\diagdown} & \xleftarrow{O_2/RH} & -CH_2CHO \; + \; \cdot CH_2CH_2- \\
\quad\;\; \diagdown OOH & &
\end{array}
$$

$$
\downarrow \Delta H/h\nu
$$

$$
-CH_2C\overset{\textstyle O}{\underset{\textstyle OH}{\diagdown}}
$$

provided a good deal of information on the mechanism of oxidative degradation of the hydrocarbon polymers.

4.1.5. The effect of chemical structure on oxidation rate

Polymers of different chemical structures vary markedly in their resistance to oxidative degradation. The reason for this will be evident from a consideration of equation (*i*) which applies to polymers undergoing oxidation in any situation where there is no limitation in oxygen supply (e.g. in an air oven at moderate temperatures) and which contain an initiator.

The rate constant $k_{4.2}$ depends on the energy of the transition state in the hydrogen abstraction reaction (scheme 4.3). In the transition state

Scheme 4.3 Transition state in the autoxidation of a hydrocarbon.

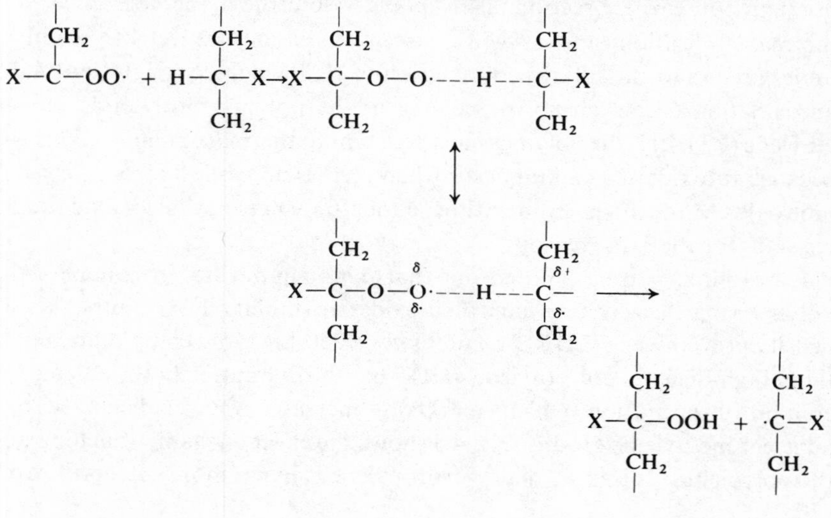

$$
\begin{array}{cccc}
\overset{\displaystyle CH_2}{\underset{|}{}} & \overset{\displaystyle CH_2}{\underset{|}{}} & \overset{\displaystyle CH_2}{\underset{|}{}} & \overset{\displaystyle CH_2}{\underset{|}{}} \\
X-\overset{|}{C}-OO\cdot \; + \; H-\overset{|}{C}-X & \rightarrow & X-\overset{|}{C}-O-O\cdots H\cdots\overset{|}{C}-X \\
\underset{|}{\overset{|}{C}H_2} & \underset{|}{\overset{|}{C}H_2} & \underset{|}{\overset{|}{C}H_2} & \underset{|}{\overset{|}{C}H_2}
\end{array}
$$

$$
\updownarrow
$$

$$
\begin{array}{cc}
\overset{\displaystyle CH_2}{\underset{|}{}} & \overset{\displaystyle CH_2}{\underset{|}{}} \\
X-\overset{|}{C}-O-\overset{\delta}{\underset{\delta\cdot}{O}}\cdots H\cdots\overset{\delta+}{\underset{\delta\cdot}{C}}- & \longrightarrow \\
\underset{|}{\overset{|}{C}H_2} & \underset{|}{\overset{|}{C}H_2}
\end{array}
$$

$$
\begin{array}{cc}
\overset{\displaystyle CH_2}{\underset{|}{}} & \overset{\displaystyle CH_2}{\underset{|}{}} \\
X-\overset{|}{C}-OOH \; + \; \cdot\overset{|}{C}-X \\
\underset{|}{\overset{|}{C}H_2} & \underset{|}{\overset{|}{C}H_2}
\end{array}
$$

the carbon atom to which the labile hydrogen is attached may assume a partial electron delocalisation ($\delta\cdot$), a partial ionic charge ($\delta+$, $\delta-$) or a combination of both depending on the nature of X. In hydrocarbon polymers in solution, the rate of oxidation increases in the series:

$$-CH_2- \; < \; -\underset{\underset{CH_3}{|}}{CH}- \; < \; -\underset{\underset{C_6H_5}{|}}{CH}- \; < \; -CH_2CH=CH- \; < \; -\underset{\underset{CH_3}{|}}{C}=CHCH_2-$$

$$\text{(PE)} \qquad \text{(PP)} \qquad \text{(PS)} \qquad \text{(PB)} \qquad \text{(PI)}$$

This is primarily due to the electron delocalising effect of the attached group but polar effects are also superimposed (CH_3 is electron-releasing).

In polymers containing hetero-atoms, polarity normally predominates. Thus, the rate of oxidation decreases in the series:

$$-CH_2- \; > \; -CH_2NHCO-CH_2- \; > \; -CH_2OCO-C_6H_4-$$

$$\text{(PE)} \qquad\qquad \text{(Nylon)} \qquad\qquad\qquad \text{(PET)}$$

$$> \; -\underset{\underset{Cl}{|}}{CH}- \; > \; -\underset{\underset{COOCH_3}{|}}{CH}- \; > \; -\underset{\underset{CH_3}{|}}{\overset{\overset{COOCH_3}{|}}{C}}- \; > \; -\underset{\underset{CN}{|}}{CH}- \gg -\overset{\overset{F}{|}}{\underset{\underset{F}{|}}{C}}-\overset{\overset{F}{|}}{\underset{\underset{F}{|}}{C}}-$$

$$\text{(PVC)} \qquad \text{(PMA)} \qquad \text{(PMMA)} \qquad \text{(PAN)} \qquad \text{(PTFE)}$$

However, this is not necessarily the order of thermal stability since as was pointed out earlier, section 2.4.1, polyacrylonitrile undergoes a facile thermal cyclisation, and as will be seen in section 4.2.1, PVC readily undergoes a radical initiated elimination of HCl which dominates normal autoxidation effects during processing of the polymer. Polytetrafluoroethylene (PTFE) is the polymer most resistant to thermal oxidation. This is because autoxidation cannot occur when there is no labile hydrogen in the molecule. Many high temperature ladder polymers are also oxidation resistant for the same reason.

It has already been pointed out that minor impurities in commercial polymers may largely determine their oxidative stability. For example, low-density polyethylene (LDPE) contains considerably more chain branching than high-density polyethylene (HDPE). Furthermore, both polymers contain unsaturation which markedly increases the oxidisability of adjacent methylenic groups. Fig. 4.3 shows the effect of chain branching in the polyolefins. It is clear that polypropylene is much more susceptible to oxidation than high-density polyethylene and low-density polyethylene lies between them.

Polystyrene represents something of an anomaly in the above generalisation. It is oxidised extremely readily in solution but in solid form it is relatively stable to thermal oxidation. Unlike the other oxidisable structures discussed above, polystyrene is below its glass transition temperature ($\approx 100\,°C$) in the solid at ambient temperature. Consequently, segmental motion of the polymer chains is not possible and a direct consequence of this is that oxygen diffusion is very much lower than it is in the polyolefins or rubbers.

The development of impact modified polystyrenes has created new problems in stabilisation. The two in most common use are high impact polystyrene (HIPS) and acrylonitrile-butadiene-styrene copolymers (ABS) which depend on the presence of polybutadiene as a separate phase in the glassy matrix. Not only can oxygen diffuse more rapidly through the rubber domains, but the latter are much more susceptible than the matrix to both thermal and photo-oxidation, as a direct result of the presence of oxidisable allylic groups in the rubber back-bone. In both polymers the styrene and/or the acrylonitrile are grafted to the polybutadiene. Photo-oxidation in

Figure 4.3. Effect of chain branching of polyolefins upon their oxidisability at 139 °C. *A* linear polyethylene (1 methyl group per 1000 carbon atoms); *B* ethylene propylene copolymer (10.7 methyl groups per 1000 carbon atoms); *C* ethylene–propylene copolymer (21.0 methyl groups per 1000 carbon atoms); *D* ethylene–propylene copolymer (35.5 methyl groups per 1000 carbon atoms); *E* polypropylene, (333 methyl groups per 1000 carbon atoms). (Reproduced by kind permission of *Atmospheric Oxidation and Antioxidants*, Elsevier, 1965, p. 275.)

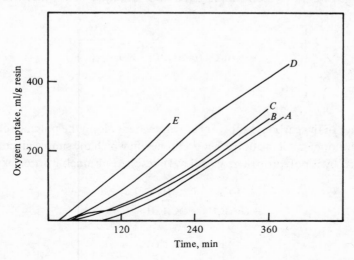

particular causes rapid hydroperoxidation of the polydiene segment (see scheme 4.4) with consequent impairment of the impact resistance of the polyblend.

Scheme 4.4 Photo-oxidation of ABS.

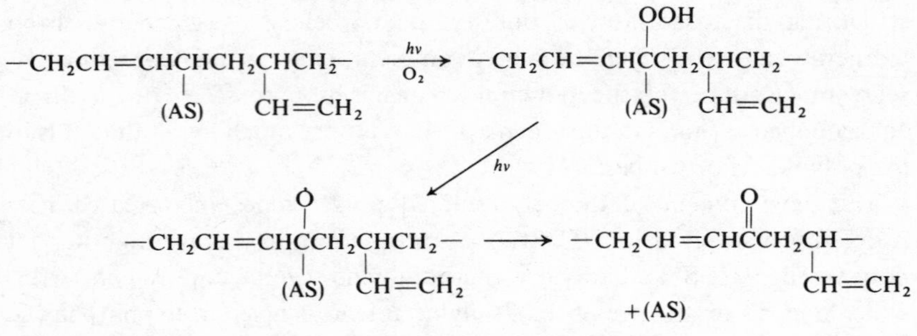

$$
-CH_2CH{=}CHCHCH_2CHCH_2{-} \xrightarrow[O_2]{h\nu}
\begin{array}{c} OOH \\ | \end{array}
-CH_2CH{=}CHCCH_2CHCH_2{-}
$$

$$
\underset{(AS)}{} \quad \underset{CH{=}CH_2}{} \qquad\qquad \underset{(AS)}{} \quad \underset{CH{=}CH_2}{}
$$

$$\searrow h\nu$$

$$
-CH_2CH{=}CHCCH_2CHCH_2{-} \longrightarrow -CH_2CH{=}CHCCH_2CH{-}
$$

$$
\underset{(AS)}{\overset{\overset{\displaystyle O}{|}}{}} \quad \underset{CH{=}CH_2}{} \qquad \underset{+(AS)}{\overset{\overset{\displaystyle O}{\|}}{}} \quad \underset{CH{=}CH_2}{}
$$

$(AS) \equiv$ acrylonitrile/styrene copolymer

Other industrially important polymers generally contain negative groups either in, or pendant to, the polymer chain and photo-oxidation does not normally present such a serious problem to durability as it does in the case of the polyolefins, the rubbers and the rubber modified plastics. However, under very aggressive environmental conditions, oxidation does occur with all these polymers. Most are subject to photo-oxidation over a period of time and the basic mechanisms outlined above have been shown to be generally applicable.

The polyurethanes are a special case. The first polyurethanes to be developed were based on simple glycols (for example V) in which the

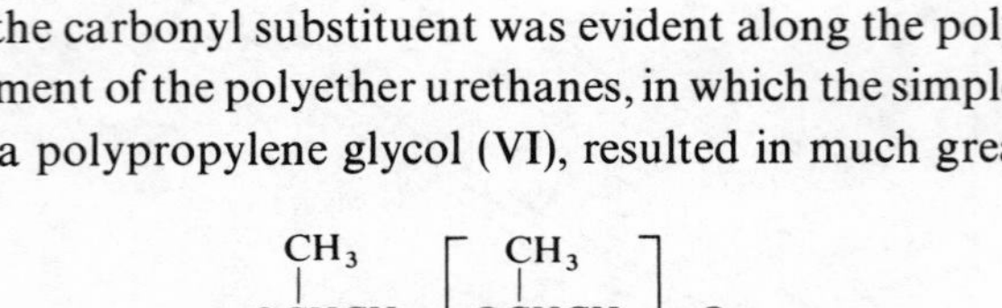

$$
\left[\underset{\underset{NH}{|}}{\overset{CH_3}{\bigcirc}}{-}NH\overset{\overset{\displaystyle O}{\|}}{C}OCH_2CH_2O\overset{\overset{\displaystyle O}{\|}}{C}{-}NH \right]_n {-}\underset{NH-}{\overset{CH_3}{\bigcirc}}
$$

V

influence of the carbonyl substituent was evident along the polymer chain. The development of the polyether urethanes, in which the simple glycol was replaced by a polypropylene glycol (VI), resulted in much greater oxygen

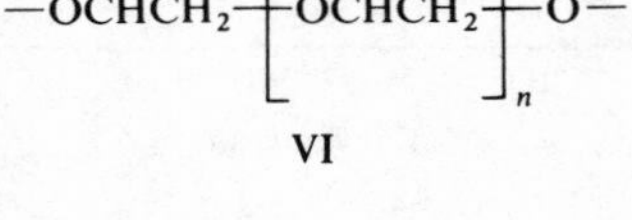

$$
\underset{}{\overset{CH_3}{-OCHCH_2}}\left[\overset{CH_3}{OCHCH_2}\right]_n{-}O{-}
$$

VI

sensitivity both during the manufacture of the polymer and during service. This results from the fact that oxygen attached to carbon activates the α-carbon atom to hydroperoxidation. The well known tendency of diethyl ether to peroxidise in the laboratory illustrates the importance of this process; in this respect an α-oxygen atom is about 20 times more effective than a methyl group and only slightly less so than an olefin group.

Consequently polyether-based polyurethanes undergo the technological phenomenon of 'scorch' during manufacture due to the oxidative sensitivity of the polyether segment and occasionally blocks of polyurethane foam have been known to undergo spontaneous ignition shortly after manufacture due to the peroxidation phenomenon.

4.1.6. Physical effects of autoxidation in polymers

The most important properties of polymers result from their high molecular weights. Their strength results from the entanglement of the polymer chains and, in semi-crystalline polymers, from the presence of crystalline regions which 'reinforce' the weaker amorphous regions. It follows from this that if molecular weight is substantially reduced by chain scission, then the important properties of polymers will be lost. Chain scission is one of the most important consequences of the autoxidation of polymers. It occurs primarily by α,β-bond scission in the alkoxyl radicals formed by the thermal or photolytic breakdown of hydroperoxides (see for example scheme 4.2). Reaction (4.11) shows the chain scission which occurs in polypropylene through breakdown of the alkoxyl radical.

$$-CH_2\overset{\overset{\displaystyle \cdot O}{|}}{C}CH_2\overset{|}{C}H- \longrightarrow -CH_2\overset{\overset{\displaystyle O}{\|}}{C} + \cdot CH_2\overset{|}{C}H- \qquad (4.11)$$
$$\overset{|}{C}H_3 \ \overset{|}{C}H_3 \overset{|}{C}H_3 \ \ \overset{|}{C}H_3$$

It is clear that since a new reactive radical is produced, this process does not affect the rate of oxidation. Reaction (4.11) is an infrequent process; the majority of the radicals hydrogen-abstract by reaction (4.8). However its practical significance cannot be over-emphasised, because only one chain scission per molecule in a polymer with a molecular mass of 100 000 destroys its technological usefulness.

This process can be followed by monitoring the change in physical behaviour of the polymer by classical physical chemical techniques of which the most frequently used is viscosity change. This may be carried out by viscometry in solution or directly on the polymer melt in a capilliary rheometer. The latter process is frequently used technologically to determine the change in melt flow index (MFI) of the polymer. Melt flow index is defined as the amount of polymer extruded through a standard orifice at a given time and temperature and there is an inverse relationship between MFI and molecular weight. Of more practical concern to the materials technologist is the change in mechanical behaviour of the polymer with time. The parameters most frequently used are tensile strength, modulus, elongation at break and impact strength. In the case of polyolefins exposed to light, tensile strength and elongation at break, initially increase and then decrease as oxidation progresses, whereas modulus changes occur in the opposite direction (see Fig. 4.4). These changes can often be related to chemical changes occurring in the polymer; for example in polyethylene, chain scission is reflected in carbonyl formation and cross-linking by gel formation, Fig. 4.4. Crystallinity frequently increases in semi-crystalline polymers due to the re-alignment of the broken chains in the crystalline domains. The effect of chemicrystallisation is to reduce the impact resistance (toughness) of the polymer.

Figure 4.4. Effect of UV irradiation on the mechanical and physical properties of low-density polyethylene during the early stages of exposure. ■ dynamic modulus; ▲ elongation at break; ▼ gel content; ● density. (Reproduced by kind permission of *J. Polym. Sci.*, Symposium No. 57, 357, 1976.)

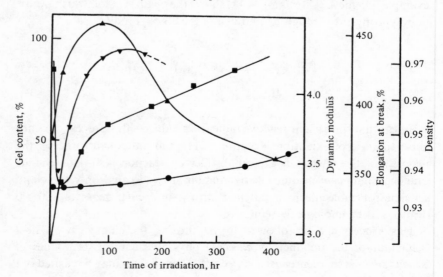

4.1.7. The effects of the physical structures of polymers on their rates of deterioration

Amorphous polymers oxidise homogeneously and uniformly throughout the bulk of the polymeric matrix. In general, polymers above their glass transition temperatures (e.g. rubbers) oxidise more rapidly than those in the glassy state (e.g. polystyrene or polymethyl methacrylate) due to the faster rate of diffusion of oxygen in the former.

The solubility of oxygen in solid polymers is much lower than it is in liquid hydrocarbons. This fact, combined with the lower rate of oxygen diffusion in polymers than in their lower molecular weight analogues, means that the alkyl to alkylperoxyl ratio in polymers may be over 100 times higher than it is in liquid hydrocarbons during autoxidation. This phenomenon makes possible the involvement of alkyl radicals in the termination step (reaction (4.4)) and in inhibition processes in polymers (see chapter 5) which would be highly unlikely in liquid hydrocarbons.

Semi-crystalline polymers are essentially two phase systems consisting of spherulitic clusters of crystals embedded in the amorphous continuum. Oxygen can diffuse readily through the amorphous regions of PE and PP but cannot penetrate the crystalline regions*, consequently most of the oxidation damage discussed in the previous sections occurs at the spherulite boundaries, thus weakening the 'adhesive' which holds the crystalline regions together. A consequence of this is that the same amount of oxidation creates much more damage in a highly crystalline polymer than it does in a less crystalline polymer of similar chemical structure. Thus, although high density polyethylene oxidises less readily than low density polyethylene due to the presence of much fewer tertiary carbon atoms, it undergoes physical degradation much more rapidly. Consequently the photo-oxidative embrittlement time for unstabilised HDPE is only about one quarter of that of unstabilised LDPE. It will be seen later that in spite of this, the crystalline polyolefins respond to antioxidants and stabilisers much more favourably than the amorphous low crystalline polymers because these additives concentrate in the amorphous phase of semi-crystalline polymers.

Similar problems are encountered in other two-phase polymers. In rubber modified polystyrene, the polystyrene phase contains occluded domains of polystyrene surrounded by a rubber-styrene block copolymer which acts as a solid phase dispersant for the rubber in the polystyrene matrix (see Fig. 4.5). However, the rubber blocks are much less oxidatively

---

* This is not however true of poly-4-methyl-pent-2-ene, where the 'open' structure of the crystalline lattice allows oxygen penetration.

stable than the polystyrene homopolymer and this leads to very rapid loss of the energy absorbing properties of the 'adhesive' between the polystyrene domains. Consequently, although the initial impact resistance of high-impact polystyrene (HIPS) is higher than PS itself, the reverse is true within a short time after exposure to the environment due to destruction of the rubber phase.

4.2. Oxidative degradation of commercial polymers

Most commercial polymers are relatively stable to oxidation as they are manufactured. For example, polyamides, polyesters, polyethylene or polypropylene as they come from the manufacturing process show only slow chemical or physical changes if stored at ambient temperatures in the dark. However, in their conversion to fabricated end products they are subjected to high temperatures and high shearing forces in the polymer melt (generally at temperatures above 150 °C and sometimes as high as 300 °C). Since it is impossible to exclude molecular oxygen completely from such procedures, important chemical modification inevitably occurs at this stage in the history of the polymer and this has a profound effect on its

Figure 4.5. Electron micrograph of photo-oxidised rubber-modified polystyrene showing the growth of cracks from the degraded rubber interphase.

1 mμ

subsequent service performance; for example at elevated temperatures (up to 150 °C) or on exposure to sunlight for relatively extended periods. It has already been seen that hydroperoxides are the primary products of thermal oxidation and these species are therefore potential initiators for subsequent oxidative degradation reactions.

4.2.1. Degradation during melt processing

The conversion of a thermoplastic polymer to a finished article normally involves heating it to the liquid state followed by extrusion through a die or into a mould. The polymer is mixed continuously by means of a screw which conveys it to the extrusion port. During this processing operation, considerable shear is applied to the viscous polymer melt which causes some of the polymer chains to undergo homolytic scission at the carbon–carbon bonds with the formation of macro-alkyl radicals. Although this is an infrequent process, the macro-alkyl radicals so produced are highly active chemical species which, like lower molecular weight radicals, initiate the radical chain reaction (4.1) and (4.2).

There are a number of quite conclusive pieces of evidence which suggest that mechanical rupture of polymer chains results in radicals. Firstly radical scavengers have a profound effect on the reaction. Thus, at ordinary temperatures no appreciable change in molecular weight occurs during mastication of natural rubber in the absence of oxygen, see Fig. 4.5. When oxygen is introduced, however, degradation is immediate and rapid. It is clear that during mastication in an inert atmosphere the primarily formed macroradicals recombine. Oxygen on the other hand reacts efficiently with these radicals, resulting in a permanent chain break. Addition of the radical acceptor, benzoquinone, has no additional effect in the presence of oxygen because oxygen is already stabilising all the radicals formed. However, in nitrogen, 1% of benzoquinone causes rapid degradation. This mechanism of stabilisation was confirmed when it was shown that in both rubber containing diphenylpicryl hydrazyl and polystyrene containing 1,1'-dinaphthyl disulphide, the amount of radical acceptor chemically incorporated into the polymer corresponded closely to the number of chain scissions measured by the decrease in molecular weight.

The weakest bonds in natural rubber are those between the individual isoprene units because the radicals so formed are highly stabilised by allylic resonance and are thus relatively unreactive in the absence of oxygen, reaction (4.12):

$$-CH_2\overset{\overset{\displaystyle CH_3}{|}}{C}=CHCH_2CH_2\overset{\overset{\displaystyle CH_3}{|}}{C}=CHCH_2- \;\rightleftharpoons$$

$$-CH_2\overset{\overset{\displaystyle CH_3}{|}}{C}=CHCH_2^{\cdot}- \;+\; {\cdot}CH_2\overset{\overset{\displaystyle CH_3}{|}}{C}=CHCH_2-$$

$$(4.12)$$

These radicals are so unreactive that they cannot take part in hydrogen abstraction processes that would be necessary to give a permanent chain break. Instead they predominantly recombine. In the case of most saturated polymers, the primarily formed radicals are very much more reactive, so that degradation occurs even in nitrogen.

A second piece of evidence in favour of scission to form macroradicals is the fact that if the polymer is degraded in the presence of a vinyl monomer and in the absence of oxygen, a block copolymer is formed. Thus, for example, the macroradical formed in the mechanodegradation of *cis*-polyisoprene in reaction (4.12) is able to initiate the polymerisation of methylmethacrylate, reaction (4.13)(*a*), to give a modified rubber. This

Figure 4.6. Effect of radical capture agents in the mastication of natural rubber. *A* in nitrogen; *B* in nitrogen + 1% benzoquinone; *C* in air. (Reproduced by kind permission of *Atmospheric Oxidation and Antioxidants*, Elsevier, 1965, p. 467.)

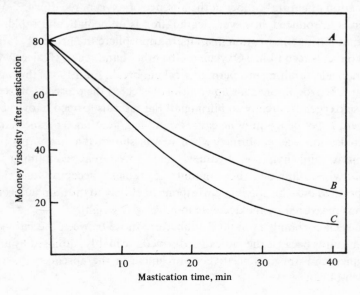

$$
\underset{\text{—CH=CH\.{C}H}_2}{\overset{\text{CH}_3}{|}} \xrightarrow[\underset{(a)}{}]{n\ \text{CH}_2=\overset{\text{CH}_3}{\overset{|}{\text{CCOOMe}}}} \underset{\text{—C=CHCH}_2}{\overset{\text{CH}_3}{|}}\!\!\left[\underset{\underset{\text{COOMe}}{|}}{\overset{\text{CH}_3}{\overset{|}{\text{CH}_2\text{C·}}}}\right]_n
$$

$$(b)\ \searrow\ \text{O}_2/\text{RH}$$

$$
\underset{\text{—C=CHCH}_2\text{OOH}}{\overset{\text{CH}_3}{|}}
\tag{4.13}
$$

process is completely inhibited by oxygen, which is able to compete with methylmethacrylate for the macro-alkyl radical and rapid degradation of the rubber occurs instead, reaction (4.13)(*b*). If two polymers are mixed together under these conditions and again in the absence of oxygen, polymer molecules which contain segments of both polymers are formed, indicating that chain scission, followed by combination of two different macro-alkyl radicals must have occurred, reaction (4.14):

$$—AAAAAAAAA—\qquad —AAAA^{\text{·}} + {}^{\text{·}}AAAAAA—$$

$$+ \qquad \rightarrow \qquad\qquad\qquad \rightarrow\ —AAAABBBBB—$$

$$—BBBBBBBBB—\qquad —BBBBB^{\text{·}} + {}^{\text{·}}BBBBBB—$$

$$\tag{4.14}$$

Thirdly, it was demonstrated that polymers masticated in the presence of the scavengers dinaphthyl disulphide and mercaptobenzthiazole labelled with radioactive ^{35}S developed built-in radioactivity.

Finally, and in many ways most conclusively, radical formation has been directly confirmed by the use of ESR spectroscopy. This technique demonstrated that the primarily formed radicals are invariably at the end of the polymer chain as required for chain scission. This is shown for poly(methyl methacrylate) in reaction (4.15)(*a*) but other radical types are formed in secondary processes, reaction (4.15)(*b*)

$$
\underset{\underset{\text{COOMe}}{\overset{|}{\text{COOMe}}}}{\overset{\text{CH}_3\ \text{CH}_3}{\overset{|\ \ \ |}{\text{—CH}_2\text{CCH}_2\text{C}—}}} \xrightarrow[\underset{(a)}{}]{\text{Shear}} \underset{\overset{|}{\text{COOMe}}}{\overset{\text{CH}_3}{\overset{|}{\text{—CH}_2\text{C·}}}}\ +\ \underset{\overset{|}{\text{COOMe}}}{\overset{\text{CH}_3}{\overset{|}{{}^{\text{·}}\text{CH}_2\text{C}—}}}
$$

$$(b)\ \Bigg\downarrow \qquad \underset{\underset{\text{COOMe}}{\overset{|}{\text{COOMe}}}}{\overset{\text{CH}_3\ \text{CH}_3}{\overset{|\ \ \ |}{—\text{CH}_2\text{CCH}_2\text{C}—}}}$$

$$
\underset{\underset{\text{COOMe}}{\overset{|}{\text{COOMe}}}}{\overset{\text{CH}_3}{\overset{|}{—\text{CH}_2\text{CCHC}—}}}\ +\ \underset{\overset{|}{\text{COOMe}}}{\overset{\text{CH}_3}{\overset{|}{\text{CH}_3\text{C}—}}}
\tag{4.15}
$$

Chemical plasticisation (mastication) of rubbers

Mechanodegradation is put to positive use during the compounding of rubber. Compounding ingredients can be incorporated into rubber only after it has been plasticised by molecular weight reduction. This is done conveniently and cheaply by making use of the chemistry described above. The rubber is normally masticated in an internal mixer or on a two-roll mill in the presence of oxygen.

An early observation by rubber technologists was that the extent of plasticisation of the rubber decreased as the temperature was increased up to a certain temperature and it then increased again at higher temperatures. This is illustrated for natural rubber in Fig. 4.7 which shows that this rubber is least susceptible to mechanodegradation at about 100 °C. The negative temperature coefficient of the reaction up to 100 °C reflects the decreasing viscosity of the rubber with increase in temperature, which in turn determines the stress on the polymer molecules and hence the rate of macro-alkyl radical formation. It will be seen later that this process is very similar to the effect of stress in vulcanised rubbers which are subject to the phenomenon of 'fatigue', which is another form of mechano-oxidation.

Figure 4.7. Temperature dependence of efficiency of mastication of rubber in oxygen. (Reproduced by kind permission of *Polymer Science*, ed. A. D. Jenkins, North Holland Publishing Company, 1972, p. 1510.)

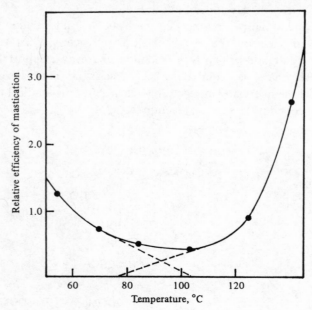

Above 100 °C hydroperoxides rather than shear play a primary role in the initiation of oxidative degradation. Under these conditions, the rate of oxidation, and hence of chain scission, is governed by the laws of chemistry rather than the laws of physics, and the degradation shows the normal positive temperature coefficient. The minimum in the efficiency curve is thus a composite of two different processes; an oxygen stabilised mechanoscission of the polymer chain below 100 °C, and hydroperoxide initiated oxidation above 100 °C.

Melt degradation of polyolefins

The reduction in molecular weight which occurs in rubbers during compounding is not a serious disadvantage when it is subsequently fabricated since compounding is followed by vulcanisation which converts the polymer to a cross-linked network. However, molecular weight change is a much more serious problem in the case of thermo-plastic polymers since, as has already been indicated, the final properties of the fabricated product depend on the molecular weight of the polymer. Furthermore, hydroperoxides and their breakdown products are a potential source of oxidative instability in the product.

Different hydrocarbon polymers behave differently during processing. Figure 4.8 shows that whereas polypropylene undergoes an increase in melt

Figure 4.8. Effect of processing of polyolefins in a shearing mixer. (*a*) polypropylene at 180 °C, closed mixer; (*b*) low density polyethylene at 150 °C, closed mixer; (*c*) polypropylene (▲) and low density polyethylene (▼) in a closed mixer after purging with argon. (Reproduced by kind permission of *Developments in Polymer Stabilisation – 5*, ed. G. Scott, App. Sci. Pub., London, 1982, p. 80.)

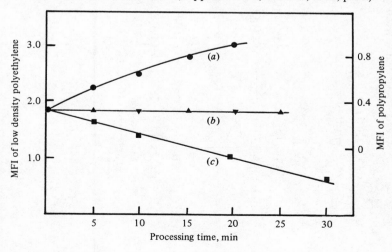

Scheme 4.5 Alternative reactions of alkyl radicals during high-temperature processing of polyolefins.

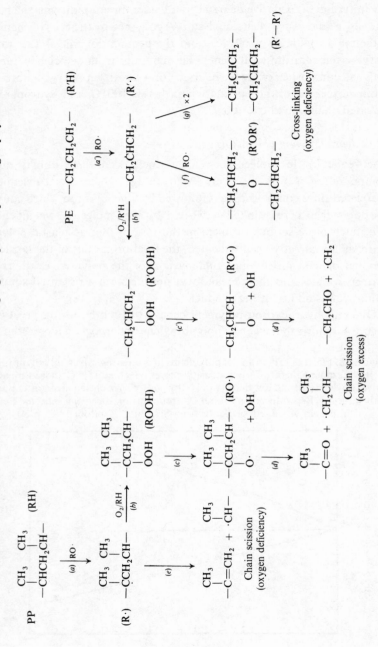

flow index (MFI) during processing in a shearing mixer, the MFI of polyethylene decreases due to cross-linking. In both cases, the initial shearing process is similar to that in rubber, see scheme 4.5. The difference lies in the subsequent reactions of the macro-alkyl radicals after depletion of oxygen which is both dissolved in the polymer and trapped between the polymer particles. Increasing the concentration of oxygen by deliberately allowing access of air leads to the predominance of chain scission in all polymers, see scheme 4.5 (reactions *d,d'*). The peroxide concentration increases with processing time in both PE and PP, but much more rapidly in an open mixer than in a closed mixer (see Fig. 4.9). In low oxygen concentrations and particularly at very high temperatures, polypropylene undergoes depolymerisation, scheme 4.5, reaction (*e*), whereas in polyethylene under similar conditions, cross-linking predominates, scheme 4.5, reactions (*f*) and (*g*).

Melt degradation of poly(vinyl chloride)

Chain scission is not the predominant chemical reaction occurring in PVC during processing. In unstabilised PVC, elimination of hydrogen chloride

Figure 4.9. Effect of processing conditions on the formation of hydroperoxides in low-density polyethylene. ■ 150 °C, open mixer; ▲ 165 °C, open mixer; ▼ 175 °C, open mixer; ● 150 °C, closed mixer. (Reproduced by kind permission of *Europ. Polym. J.*, **13**, p. 731, 1977.)

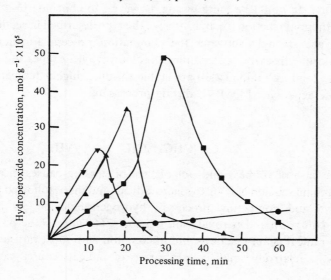

by the well-known 'unzipping' reaction, scheme 4.6, reactions (*b*), (*c*) and (*d*), is much more important since it leads to quite intense discolouration of

Scheme 4.6 Mechano-chemical reactions of poly(vinyl chloride) during processing.

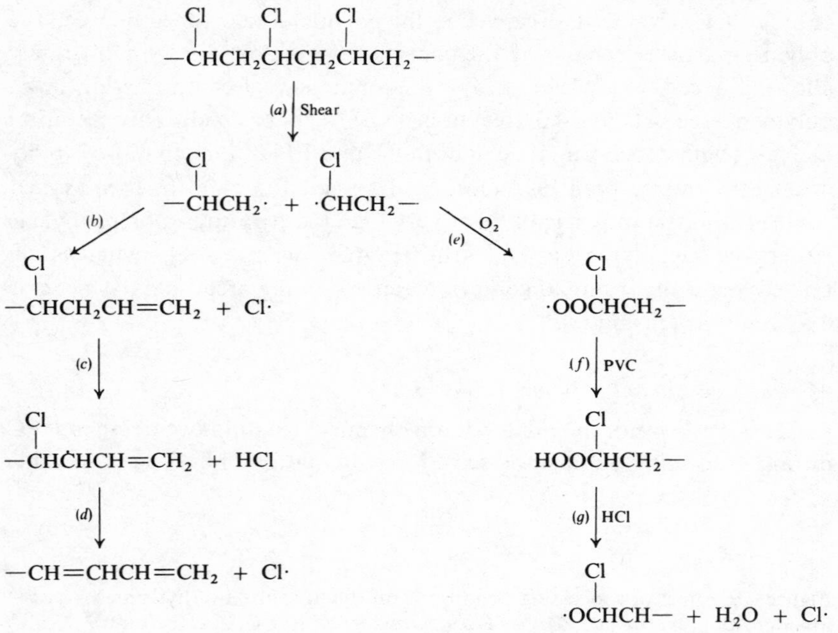

the polymer. It would be quite wrong, however, to conclude that this is simply a thermal reaction (i.e. pyrolysis) of the type described in section 2.4, since oxygen strongly activates the elimination process. Furthermore, model chlorohydrocarbons containing the chloroethylene repeating unit of PVC (e.g. 2-chlorobutane (VII)) are stable to much higher temperatures than those experienced by PVC during processing.

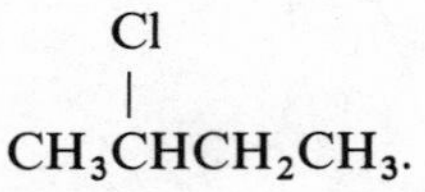

$$\underset{\qquad}{\overset{\text{Cl}}{\underset{|}{\text{CH}_3\text{CHCH}_2\text{CH}_3.}}} \qquad \text{(VII)}$$

As in the case of rubbers and polyolefins, processing causes mechano-chemical chain scission of PVC to macroradicals and this is followed by two alternative and competing processes, hydrogen chloride elimination (scheme 4.6(*b*)–(*d*)) and peroxidation (scheme 4.6(*e*),(*f*)). These processes are both radical chain reactions and in common with most radical chain reactions, a relatively small concentration of radicals causes extensive chemical change; in this case the formation of conjugated double bonds in the polymer backbone.

However, this is not the only consequence of processing since the HCl produced itself has an adverse effect on the thermal stability of PVC, particularly when a small amount of oxygen is present. It has been found that anhydrous hydrogen chloride induces the homolysis of hydroperoxides (scheme 4.6, reaction (*g*)), thus producing more free radicals which increase the rate of the radical chain reaction. It can be seen from figure 4.10 that formation of olefinic unsaturation and hydroperoxides in unstabilised PVC are closely associated with the initial period when the viscosity of the polymer and hence the applied torque in the mixer are high.

It will be seen in section 4.2.3 that hydroperoxides produced in PVC during processing are the cause of subsequent instability on exposure to light.

4.2.2. Degradation at high temperatures during service

The temperature at which most thermoplastic polymers are used is limited by the fact that they undergo dimensional changes (creep) at high temperatures. With the trend towards fibre reinforcement however, this

Figure 4.10. Relationship between applied torque (●) and the appearance of unsaturation (■), peroxides (▲) and gel (▼) during the processing of unstabilised poly(vinyl chloride).

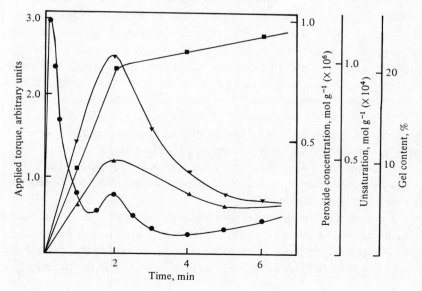

situation is rapidly changing and oxidative stability of some plastics up to temperatures of 120 °C for considerable lengths of time may be demanded in some end uses (e.g. in the motor car engine).

Vulcanised rubbers on the other hand can be and are used at high temperatures because of their cross-linked structure. Although rubbers have good short-term dimensional stability, unless effectively stabilised they undergo long-term dimensional changes (creep and permanent set) in the presence of oxygen. Rubbers based on diene monomers (cis-polyisoprene, cis-polybutadiene, styrene-butadiene and nitrile-butadiene) are all highly susceptible to oxidative degradation due to the presence of an allylic group in the polymer backbone. A repeating double bond in a hydrocarbon chain such as that in polyisoprene increases the rate of oxidation about 30 times relative to a fully saturated chain.

Change of tensile strength with time in an air oven is the most widely used technological test used to follow the degradation of a rubber network but the relaxation of stress in a sample held under restraint is a complementary test which provides a valuable insight into the chemical changes occurring.

Change in stress with time is related to the number of network chains supporting the stress by the expression

$$f/f_o = N/N_o$$

where f and N are the stress and number of network chains after a period of time and f_o and N_o are the same parameters at zero time. Stress relaxation can be carried out in two ways, either with the rubber under continuous stress or at intervals while the rubber is aged in the unstressed condition. The first measures chain scission only, whereas the second measures the resultant effect of opposed chain scission and cross-linking reactions.

Natural rubber (NR) vulcanisates undergo molecular weight reduction (stress relaxation) in both tests, whereas styrene–butadiene rubbers (SBR) undergo chain scission under continuous stress relaxation and cross-linking when the stress is measured intermittently at 130 °C (Fig. 4.11). This difference in behaviour of NR and SBR results from the presence of pendant vinyl groups in SBR which are present in all polymers containing butadiene as a comonomer. Vinyl is much more susceptible to attack by radicals than vinylene which undergoes mainly hydrogen abstraction and subsequent chain scission (see scheme 4.7).

Cross-linking is an important problem in nitrile butadiene rubber (NBR) which, because of its oil resistance, is frequently used in engine seals and gaskets at temperatures as high as 130 °C. Cross-linking caused by oxidation leads to increase of modulus and hardening of the rubber, thus reducing its ability to function as a seal. The relative contribution of chain scission and cross-linking during accelerated ageing of rubbers depends on the temperature of test and the availability of oxygen in the system. NR undergoes predominant chain scission above 80 °C but below 80 °C it cross-links (Fig. 4.12). This illustrates the change in relative importance of cross-

Figure 4.11. Continuous (——) and intermittent (– – –) stress relaxation of SBR vulcanisates (50% elongation) at 130 °C. *A,A'*, tyre tread vulcanisate; *B,B'*, gum vulcanisate (without carbon black). (Reproduced by kind permission of *Atmospheric Oxidation and Antioxidants*, Elsevier, 1965, p. 413.)

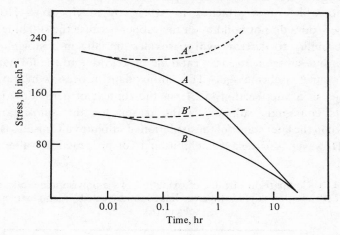

Figure 4.12. Effect of temperature on the physical change which occurs during the oxidation of a natural rubber vulcanisate. (Reproduced by kind permission of *Atmospheric Oxidation and Antioxidants*, Elsevier, 1965, p. 424.)

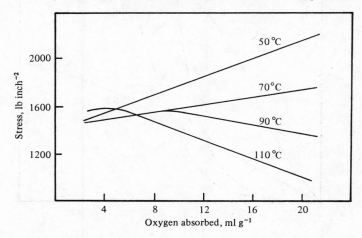

linking and chain scission with temperature change and indicates the difficulty of predicting physical changes in service from oxidation at a single temperature. In general it may be concluded that the latter is preferred over the former the higher the temperature.

Most commercial rubber products are made by sulphur vulcanisation and the introduction of a sulphur cross-link further activates the rubber molecule. This is shown in Fig. 4.13 which compares the behaviour of a peroxide cross-linked cis-polyisoprene rubber (containing only carbon–carbon cross-links) with a polysulphide cross-linked rubber in an oxygen absorption test at 80 °C. Sulphur, like oxygen, activates the carbon atom to which it is attached to attack by alkylperoxyl. However, hydroperoxides do not build up in the rubber because the sulphide group has the ability to destroy hydroperoxides, initially in a stoichiometric reaction but subsequently in a catalytic process due to the formation of sulphur acids (scheme 4.8). This antioxidant process which will be discussed in a later section (5.2.3), in the context of peroxidolytic anti-oxidants containing sulphur, is the reason for the auto-retardation observed in the later stages of the oxidation of sulphur vulcanisates (see Fig. 4.13). However, this delayed inhibition is of no practical value to the

Figure 4.13. Comparison of rates of oxidation of cis-polyisoprene vulcanisates; peroxide (P) and polysulphide (S) cross-linked; unextracted (U) and extracted with solvent (E).

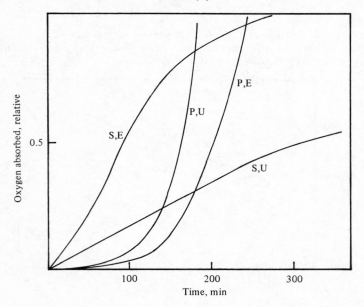

Scheme 4.7 Alternative reactions of unsaturated groups in polybutadiene rubbers.

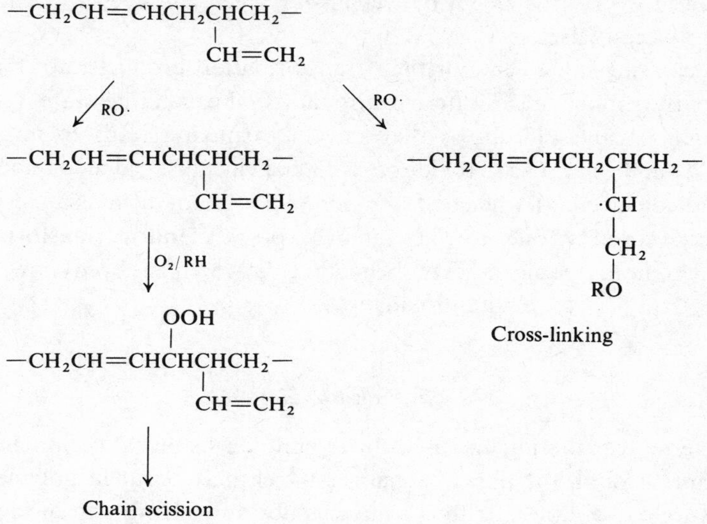

$$-CH_2CH=CHCH_2CHCH_2-$$
$$\overset{|}{CH=CH_2}$$

RO· (left branch) RO· (right branch)

Left:
$$-CH_2CH=CH\dot{C}HCHCH_2-$$
$$\overset{|}{CH=CH_2}$$

O_2/RH

$$\overset{OOH}{\overset{|}{-CH_2CH=CHCHCHCH_2-}}$$
$$\overset{|}{CH=CH_2}$$

Chain scission

Right:
$$-CH_2CH=CHCH_2CHCH_2-$$
$$\overset{|}{\cdot CH}$$
$$\overset{|}{CH_2}$$
$$\overset{|}{RO}$$

Cross-linking

polymer technologist since it is a concomitant of chain scission (scheme 4.8, reactions (*b*) and (*c*)).

Scheme 4.8 Oxidative cross-link scission in a rubber vulcanisate.

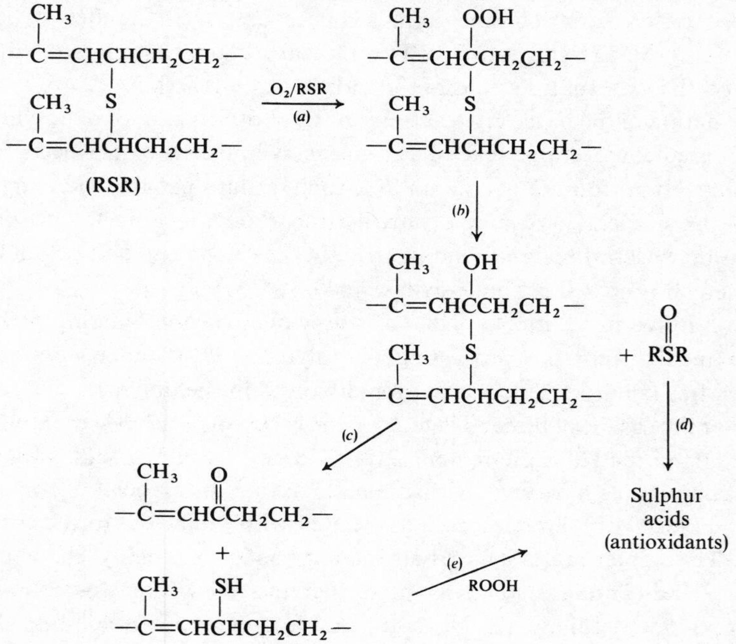

Left (RSR):
$$\overset{CH_3}{\overset{|}{-C=CHCHCH_2CH_2-}}$$
$$\overset{CH_3}{\overset{|}{-C=CHCHCH_2CH_2-}} \quad \overset{S}{\overset{|}{}}$$
(RSR)

O_2/RSR (*a*)

Right:
$$\overset{CH_3}{\overset{|}{-C=CH}}\overset{OOH}{\overset{|}{C}}CH_2CH_2-$$
$$\overset{CH_3}{\overset{|}{-C=CHCHCH_2CH_2-}} \quad \overset{S}{\overset{|}{}}$$

(*b*)

$$\overset{CH_3}{\overset{|}{-C=CH}}\overset{OH}{\overset{|}{C}}CH_2CH_2-$$
$$\overset{CH_3}{\overset{|}{-C=CHCHCH_2CH_2-}} \quad \overset{S}{\overset{|}{}}$$

$$+ \ \overset{O}{\overset{\parallel}{RSR}}$$

(*c*) (*d*)

(*c*) branch:
$$\overset{CH_3}{\overset{|}{-C=CH}}\overset{O}{\overset{\parallel}{C}}CH_2CH_2-$$
$$+$$
$$\overset{CH_3}{\overset{|}{-C=CHCHCH_2CH_2-}}\quad\overset{SH}{\overset{|}{}}$$

(*e*) ROOH

(*d*) branch:
Sulphur acids (antioxidants)

Consequently, the tensile strength of vulcanised rubber is completely destroyed by as little as 3% by weight of oxygen and even 1% renders it technologically useless.

Vulcanising ingredients or their transformation products, not forming part of the rubber vulcanisate, can also modify the oxidative stability of the product. Figure 4.13 shows that solvent extraction leads to improved stability in the case of a peroxide cross-linked vulcanisate due to removal of unreacted peroxides, whereas extraction of a sulphur vulcanisate results in reduced stability due to the removal of accelerator transformation products, notably zinc mercaptobenzothiazole which has been shown to be a very powerful thermal antioxidant.

4.2.3. Photo-oxidation

The term 'weathering' is used by technologists as a comprehensive description of all the possible changes which may occur in polymers on exposure out-of-doors. It thus embraces not only changes in mechanical behaviour (tensile strength, impact strength, etc.) but also aesthetically undesirable changes such as discolouration, loss of gloss, etc. Although moisture and humidity can have secondary effects in weathering, the primary process occurring is photo-oxidation or perhaps more accurately, photo-initiated oxidation, since, as has already been seen, the effect of light is primarily on the generation of free radicals. Light has relatively little effect on the propagating steps of the radical chain reaction.

The nature of the initiating reactions in the photo-oxidation of polymers has aroused considerable scientific controversy in recent years since 'pure' polymers do not normally contain functional groups capable of acting as sensitising species. For example, pure hydrocarbons show no UV absorption in the spectral region found in sunlight (i.e. $>285\,nm$) and yet all the commercial polyolefins photo-oxidise readily.

It was noted in section 4.2.1 that a variety of oxygen-containing groups are formed during the processing of polymers even under nominally oxygen-free conditions, due to oxygen dissolved in the polymer.

By far the easiest of these to detect are the group of carbonyl compounds absorbing in the IR in the region 1710–1735 cm^{-1} (see Fig. 4.2). Some of these compounds have strong absorbances in the sun's special region and show characteristic luminescence associated with excitation to the triplet state. The triplet states of carbonyl compounds are highly chemically reactive species and can undergo three different kinds of transformation in polymers. Two of these, the Norrish I and II processes were discussed in chapter 2, section 3.2, but much more important, from the point of view of photo-oxidation, is their reduction by hydrogen abstraction from the polymer, reaction (4.16). The alkyl radicals produced can react with oxygen

$$\begin{array}{c} R' \\ \diagdown \\ \diagup \\ R'' \end{array}\!\!C\!-\!\dot{O} + RH \longrightarrow \begin{array}{c} R' \\ \diagdown \\ \diagup \\ R'' \end{array}\!\!C\!-\!OH + R\cdot \qquad (4.16)$$

as in reaction (4.1), the first step of the chain reaction. Although the Norrish I process (reaction (3.3)) is potentially capable of initiating the autoxidation of the polymer, the two radicals are formed in a 'cage' due to the high viscosity of the polymer in the solid state and the evidence suggests that Norrish I initiation of autoxidation does not occur to any significant extent during the early stages of photo-oxidation of polyolefins. Instead, the Norrish II process, 3.4, leads to photolysis of the polymer with associated chain scission but because it does not give rise to free radicals, it does not catalyse oxidative degradation. Polymers have to be processed very severely before appreciable quantities of carbonyl compounds appear since, as was seen in section 4.2.1, they are formed by thermolysis or photolysis of hydroperoxides. Polyolefins which have been subjected to very severe processing in order to produce carbonyl compounds in the polymer chain photo-oxidise much more rapidly than mildly processed polymers (see Fig. 4.14). However, thermolysis of a severely processed sample in argon, which

Figure 4.14. Effect of processing on the rate of carbonyl formation during the photo-oxidation of polypropylene. Numbers on curves are processing time in minutes. *A* indicates heated in argon for 50 hours at 110 °C before UV exposure. (Reproduced by kind permission of *Polymer*, **18**, p. 98, 1977.)

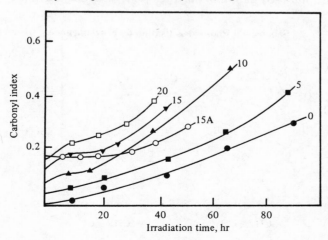

effectively removes hydroperoxides, increases the photostability of the polymer to that of a control which has not been subjected to oxidative degradation during processing (compare curves 15 and 15A in Fig. 4.14). PVC has been found to behave in an analogous manner and in both polyolefins and PVC there is a 'linear' relationship between the initial concentration of hydroperoxides in polymers processed with increasing severity and their photo-oxidation rate. Furthermore, a plot of $[ROOH]^{1/2}$ is linear with time during the initial stages of photo-oxidation, which by comparison with the rate equation (1), clearly establishes that hydroperoxide photolysis (reaction (4.7)) is the initiating step. Carbonyl compounds absorb light very effectively (i.e. they have a high molar absorptivity) whereas hydroperoxides absorb light inefficiently. However, whereas the former have a low quantum efficiency for radical formation by the Norrish I process, hydroperoxides produce radicals with a quantum efficiency of approximately I due to the ready diffusion of the hydroxyl radical from the site of the reaction. This factor rather than the amount of light absorbed determines the relative efficiency of the two species as photo-initiators.

Aromatic ketones have a higher quantum efficiency for radical formation than aliphatic ketones and well known photosensitisers such as benzophenone (IX) may be deliberately added to polymers in order to increase their rate of photo-oxidation. This kind of sensitisation process leads to the rapid photodegradation of polypropylene but it is ineffective in polyethylene. This appears to be due to the fact that the diphenylhydroxy-methyl radical (IX) produced (scheme 4.9) is a relatively effective inhibitor for further photo-oxidation in the latter case, whereas it does not appear to inhibit the photo-oxidation of polypropylene, probably for steric reasons.

Scheme 4.9 Photo-sensitisation by benzophenone.

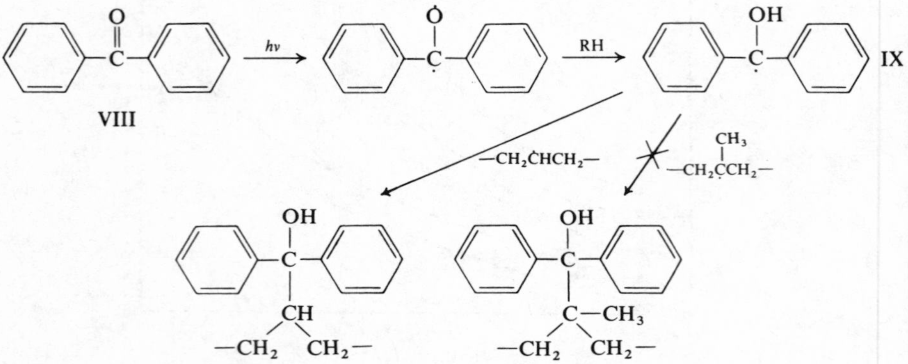

Polystyrene appears to be very strongly sensitised by its own ketonic degradation products. The primary point of attack of oxygen during processing is in this case on the α-carbon atom attached to the aromatic ring. Breakdown of the initially formed hydroperoxide leads to the formation of an alkaryl ketone (see scheme 4.10) which like benzophenone, is a powerful photo-sensitiser.

A wide variety of agents external to the polymer chain have been invoked to explain the sensitivity of polymers to photo-oxidation under specific conditions. These include polycyclic hydrocarbons present in soot in the industrial environment, atmospheric pollutants such as NO_2, SO_2 and O_3. These may all contribute to sensitisation in industrial environments (see section 7.1), but compared with the chemical changes introduced into polymers during processing, they probably have only peripheral activity in normal environments.

Scheme 4.10 Photo-oxidation of polystyrene.

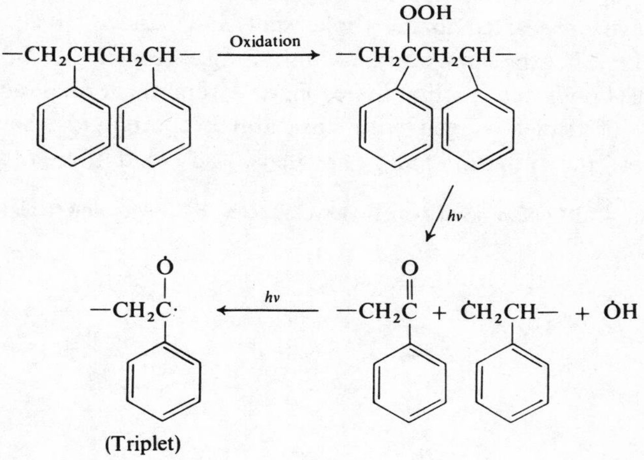

4.2.4. Sensitisation by pigments

Plastics are normally coloured by pigmentation and fibres, either by pigmentation or by dyeing. Both classes of colourants can act either as stabilisers or sensitisers for polymers. The stabilising action of pigments is generally associated with their ability to physically screen out the incident radiation. The most important example of the photo-active additives is titanium dioxide, a widely used white pigment. TiO_2 can exist in two forms, anatase, which is a photosensitiser and rutile which is essentially inert in the form in which it is used as a commercial pigment. Anatase absorbs light up to 400 nm and is readily excited to a species that in the presence of oxygen can lead to the rapid formation of hydroperoxide. Two alternative mechanisms have been invoked (see scheme 4.11). The first (*a*) involves the formation of a radical species and the second (*b*) the formation of singlet oxygen. The former

Scheme 4.11 Photo-sensitising action of TiO$_2$.

$$TiO_2 \longrightarrow [TiO_2]^* \xrightarrow{\ ^3O_2\ } [Ti^+O_2\,O_2^-] \begin{array}{c} \overset{(b)}{\nearrow} TiO_2 + {}^1O_2 \\[4pt] \underset{RH}{\overset{(a)}{\searrow}} TiO_2 + HOO\cdot + R\cdot \end{array}$$

is probably more important in polyolefins, since singlet oxygen is not reactive towards saturated hydrocarbons.

Singlet oxygen ($^1O_2$) is, however, highly reactive towards unsaturated polymers (e.g. diene-based rubbers) and hydroperoxides are formed rapidly. A number of dyestuffs, notably rhodamine, fluorescene and chlorophyll are excited to the triplet state and hand on their energy to oxygen in rather the same way as anatase. Singlet oxygen so produced adds to double bonds in a reaction quite different from the normal radical chain reaction of triplet oxygen with unsaturated compounds. Scheme 4.12 compares the normal and photosensitised oxidation of 1-methyl

Scheme 4.12 Oxidation of 1-methyl cyclohexene by singlet and triplet oxygen.

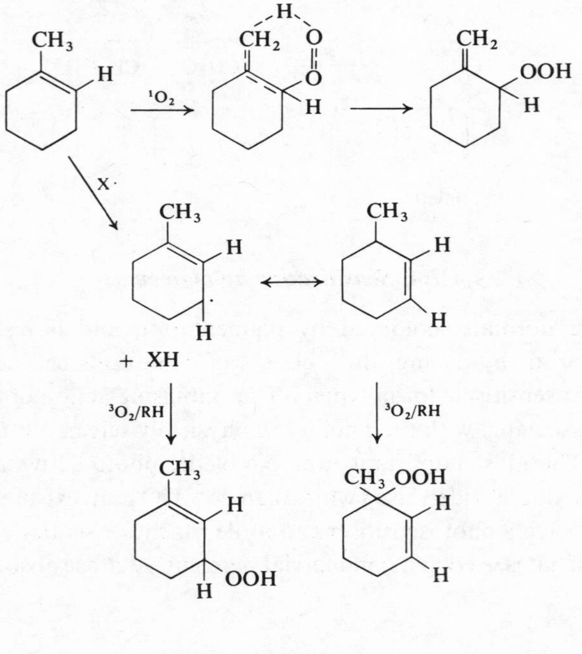

cyclohexene and it can be seen that the main difference between the two processes is the formation of an exocyclic double bond in the case of singlet oxygen attack. However, the hydroperoxides produced from singlet oxygen attack have exactly the same properties as those produced by normal autoxidation and effectively initiate photo-oxidation.

4.2.5. Mechano-oxidation

Rubbers are frequently used under conditions of high stress. In the motor car tyre for example, the tyre tread and sidewall are continually subjected to cyclical stresses during normal operation. This leads to a much faster rate of degradation than occurs in rubbers in the unstressed condition. The source of this kind of degradation is mechanical shear. The presence of the cross-link restricts spatial migration of polymer molecules and segments of the polymer are statistically subjected to much higher stresses than others. This can lead to mechanical scission of the chains or the associated cross-links. This process is essentially similar to that which occurs during the processing of polymers in a shearing mixer (see section 4.2.1) and the result is a much faster rate of initiation of the normal autoxidation process.

By analogy with the human condition, this accelerated bond scission is known as fatigue and it constitutes by far the most serious threat to the lifetime of rubbers used in mechanical goods and in particular it is a potential source of danger in tyres due to the possibility of premature failure. This phenomenon is exacerbated by the presence of ozone in the atmosphere. The problem is particularly severe in highly polluted industrial atmospheres, where photochemical reactions in the upper atmosphere give rise to substantial ozone concentrations at ground level (see section 7.1).

Ozone leads to crack formation in rubber in the stressed state. By contrast, no physical changes can normally be observed in the surface of unstressed rubber, although there is chemical evidence that ozone initially reacts just as readily with unstressed as with stressed rubber.

These phenomena can be explained in terms of the known chemistry of ozonolysis (see scheme 4.13). Initial attack of ozone occurs in both stressed and unstressed rubber to give a molozonide which is unstable and dissociates to give a zwitterion and a carbonyl compound. In the absence of stress, further combination of the two species occurs in a cage reaction to give a stable iso-ozonide but in the stressed condition, this reaction cannot occur due to separation of the chain ends. The zwitterion subsequently reacts with another identical species or with a carbonyl compound

Scheme 4.13 Chemistry of ozonolysis of rubber

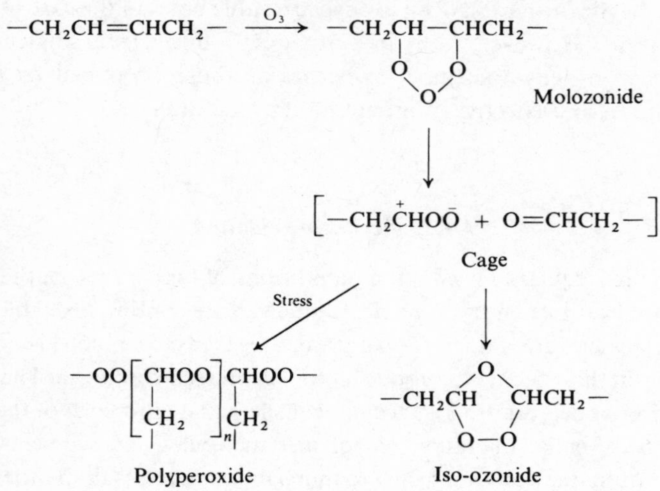

$$-CH_2CH{=}CHCH_2- \xrightarrow{O_3} -CH_2CH{-}CHCH_2-$$

Molozonide

$$\left[-CH_2\overset{+}{C}HO\bar{O} + O{=}CHCH_2-\right]$$

Cage

Stress

$$-OO\left[\begin{array}{c}CHOO\\|\\CH_2\\|\end{array}\right]_n\begin{array}{c}CHOO-\\|\\CH_2\\|\end{array}$$

Polyperoxide

$$-CH_2CH \quad CHCH_2-$$

Iso-ozonide

elsewhere in the rubber. The second process leads to a weakening of the rubber structure and incipient crack formation. If the rubber is being subjected to dynamic stresses, the development of small cracks accelerates the process of fatigue since the stresses at the tip of a growing crack are much greater than on a plane surface. Dynamic ozone cracking is, therefore, a particularly severe problem to which the tyre technologist has found empirical solutions. This will be discussed in the next chapter.

Suggested further reading

1. G. Scott, *Atmospheric Oxidation and Antioxidants*, Elsevier, 1965.
2. W. L. Hawkins (ed.), Polymer Stabilisation, Wiley & Sons, 1967.
3. B. Ranby and J. F. Rabek, *Photo-degradation, Photo-oxidation and Photo-stabilisation of Polymers*, Wiley & Sons, 1975.
4. E. T. Denisov, Role of Alkyl Radicals in Polymer Oxidation and Stabilisation, *Developments in Polymer Stabilisation-5*, ed. G. Scott, App. Sci. Pub., London, 1982, chapter 2.
5. G. Scott, The Role of Peroxides in the Photo-Degradation of Polymers, *Developments in Polymer Degradation-1*, ed. N. Grassie, App. Sci. Pub., London, 1979, chapter 7.
6. N. S. Allan and J. F. McKellar, *Photo-Chemistry of Man-made Polymers*, App. Sci. Pub., London, 1980.
7. R. Arnaud and J. Lemaire, Photo-catalytic Oxidation of Polyolefins, *Developments in Polymer Degradation-2*, ed. N. Grassie, App. Sci. Pub., London, 1979, chapter 6.
8. J. E. Stuckey, High Temperature Stability of Rubber Vulcanisates, *Developments in Polymer Stabilisation-1*, ed. G. Scott, App. Sci. Pub., London, 1979, chapter 3.
9. D. J. Carlsson, A. Garton and D. M. Wiles, The Photo-Stabilisation of Polyolefins, *Developments in Polymer Stabilisation-1*, ed. G. Scott, App. Sci. Pub., London, 1979, chapter 7.

5

Antioxidants and stabilisers

5.1. Mechanisms of antioxidant action

Antioxidants are inhibitors for autoxidation. In the present chapter the term will be used to include heat stabilisers, melt stabilisers, light (UV) stabilisers, antifatigue agents and antiozonants. All these agents interfere with the free radical reactions that lead to the incorporation of oxygen into macromolecules.

The autoxidation chain mechanism discussed in chapter 4 is summarised in scheme 5.1. It involves two interacting cyclical processes. The first cycle is the alkyl/alkylperoxyl chain reaction and the second involves the homolysis of hydroperoxides which feeds the chain reaction with new

Scheme 5.1 Mechanisms of antioxidant action.

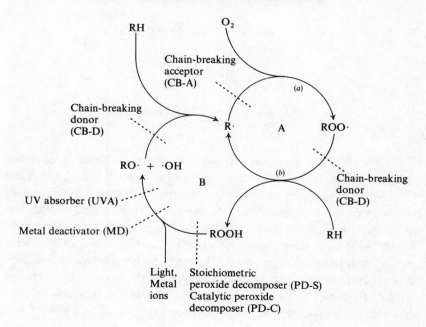

radicals. In the stationary state, the rate of formation of hydroperoxides by the A cycle is equal to their rate of decomposition by the B cycle and the rate of oxidation is constant.

The purpose of adding an inhibitor is to prevent this stage being reached by inhibiting or retarding the formation of hydroperoxides for as long as possible. Scheme 5.1 also indicates the points at which anti-oxidants can in principle operate.

5.1.1. Chain-breaking antioxidants

The primary cycle can be interrupted at two points. In the presence of an oxidising agent the alkyl radical can be removed to give a carbonium ion and subsequent inert reaction products such as olefins by elimination of a proton. The alkylperoxyl radical can be reduced to give a hydroperoxide. Antioxidants acting by these two mechanisms are electron acceptors (oxidising agents) and electron donors (reducing agents) respectively, and will be designated as CB-A and CB-D antioxidants for convenience in this chapter (see scheme 5.2).

Scheme 5.2 Kinetic chain-breaking mechanisms.

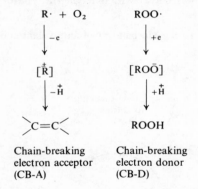

If both alkyl and alkylperoxyl radicals are present in an autoxidising system in comparable concentration at the same time, then both CB-A and CB-D mechanisms may operate simultaneously. Moreover, if the oxidised and reduced forms of an antioxidant form a reversible redox couple, then more than one kinetic chain may be broken per antioxidant molecule. Under these favourable conditions catalytic inhibition may occur and is generally represented in scheme 5.3. It can be seen that the catalytic anti-oxidant cycle in scheme 5.3 is in competition with the catalytic oxidation cycle A of scheme 5.1. It was seen earlier that in most autoxidations the rate-determining step in scheme 5.1 is reaction (*b*). Consequently the CB-A process is scheme 5.3 will only operate effectively if the oxygen concentration is low at the site of oxidation. If it is not, then oxygen can successfully compete with the oxidised form of the antioxidant for the alkyl radical, so that only the CB-D mechanism can terminate chain propagating radicals efficiently.

Scheme 5.3 General catalytic chain-breaking mechanism of antioxidant action.

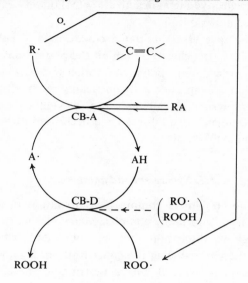

5.1.2. Preventive antioxidants

Antioxidants acting on the B cycle are 'preventive' since their primary function is to interfere with the generation of free radicals which feed back into the A cycle. The most important preventive mechanism, both theoretically and practically, is hydroperoxide decomposition (PD) in a process which does not involve the formation of free radicals. Peroxidolytic antioxidants fall into two classes; stoichiometric peroxide decomposers (PD-S), such as phosphite esters which are reagents for the reduction of hydroperoxides to alcohols, and catalytic peroxide decomposers (PD-C). A variety of sulphur compounds which are of considerable importance as commercial anti-oxidants fall into the latter class and their mechanisms have been extensively studied in recent years.

Although complete destruction of hydroperoxides from the manufacture of the polymer artifact through its service life is the ideal to be aimed at, other agents which slow down their decomposition may act as retarders of peroxide initiated oxidation. As observed in chapter 4, many transition metal ions are effective catalysts for the homolytic decomposition of hydroperoxides. They are therefore effective pro-oxidants and their activity depends on the availability of oxidised and reduced states of comparable stability. Complexing agents which have the ability to co-ordinate the vacant orbitals of transition metal ions to their maximum co-ordination number and thus inhibit the co-ordination of hydroperoxides to the metal ions are therefore effective metal deactivators (MD) and hence inhibitors for metal catalysed autoxidation.

Similarly, in the case of UV-initiated oxidation, UV light absorbers (UVA), screens, filters, etc, effectively inhibit the photo-excitation of light absorbing species present in the polymer. Another way in which preventive antioxidants may in principle act, is by quenching of photo-excited states of chromophores present in polymers (e.g. carbonyl compounds). In practice, many commercial stabilisers have been found to operate by a combination of these mechanisms.

5.1.3. Synergism and antagonism

Most stabilisers for polymers contain a combination of anti-oxidants acting by different and normally complementary mechanisms. It will be clear from the above discussion that an antioxidant which destroys hydroperoxides, thereby reducing the concentration of radicals in the A cycle will in consequence slow down the destruction of a chain-breaking antioxidant. By the same token, an effective chain-breaking antioxidant reduces the amount of hydroperoxide formed in an autoxidising system and hence protects a peroxide decomposer. This co-operative interaction, which is commonly called synergism, leads to an overall antioxidant effect which is greater than the sum of the individual effects and very often to an effectiveness much greater than can be achieved by either antioxidant alone even at much higher concentrations. The phenomenon is therefore of considerable practical and theoretical significance.

Occasionally the reverse phenomenon is observed; that is, two anti-oxidants interact to decrease the sum of their individual effects. This is described as antagonism.

5.2. Chain-breaking (CB) antioxidants

5.2.1. The chain-breaking donor (CB-D) mechanism

The earliest antioxidants to be investigated fall into this class. During the early development of rubber, it was found that some vulcanising agents led to superior 'ageing' performance of the finished product. These were identified as arylamines, and of these, diphenylamine itself (I, $R_1 = R_2 = H$) was one of the earliest antioxidants to be developed commercially.

$$I(a)\ R_1 = R_2 = tOct,\ DODPA$$
$$I(b)\ R_1 = H,\ R_2 = NHisoPr,\ IPPD$$
$$I(c)\ R_1 = H,\ R_2 = NHPh,\ DPPD.$$

I

Diphenylamine is too volatile to be used in modern rubber technology but higher molecular weight alkylated derivatives such as octylated diphenylamine (Ia) are commercial antioxidants for rubbers used at high temperatures. The 4-aminodiphenylamines (e.g. N-isopropyl-N' phenyl-p-phenylene diamine (IPPD) I(b) and diphenyl-p-phenylene diamine (DPPD) I(c) are important antidegradants for tyres. The uses and mechanisms of the diphenylamine antioxidants will be discussed later, but the chemistry of their oxidative transformations is summarised in scheme 5.4. The initially formed highly-reactive aminyl radical (IIa) reacts in two different ways to give products which themselves have antioxidant activity. The first group of products are oligomers formed by dimerisation and subsequently polymerisation through the carbon form of the aminyl radical (IIb) to give a dimeric product (III) which is just as effective as an antioxidant as the initial arylamine but it has the additional advantage of higher molecular weight, and hence, lower volatility. The second type of product formed through a second CB-D step is the nitroxyl radical (IV). It will be seen later that this species is a very important CB-A antioxidant.

A disadvantage of the arylamines is that they cause considerable discolouration of the polymers to which they are added. This is due to the formation of extensively conjugated quinonoid oxidation products such as V (scheme 5.4). Non-staining antioxidants based on substituted phenols were originally developed to satisfy the need to have white or pastel-tinted rubber products. The simplest, most important member of this class which is used as an antioxidant in other technologies (e.g. foodstuffs) is the 'hindered phenol' 2,6-di-tert-butyl phenol, VI, normally referred to as butylated hydroxytoluene (BHT).

VI, BHT

Scheme 5.4 Oxidation transformation of the diphenylamines.

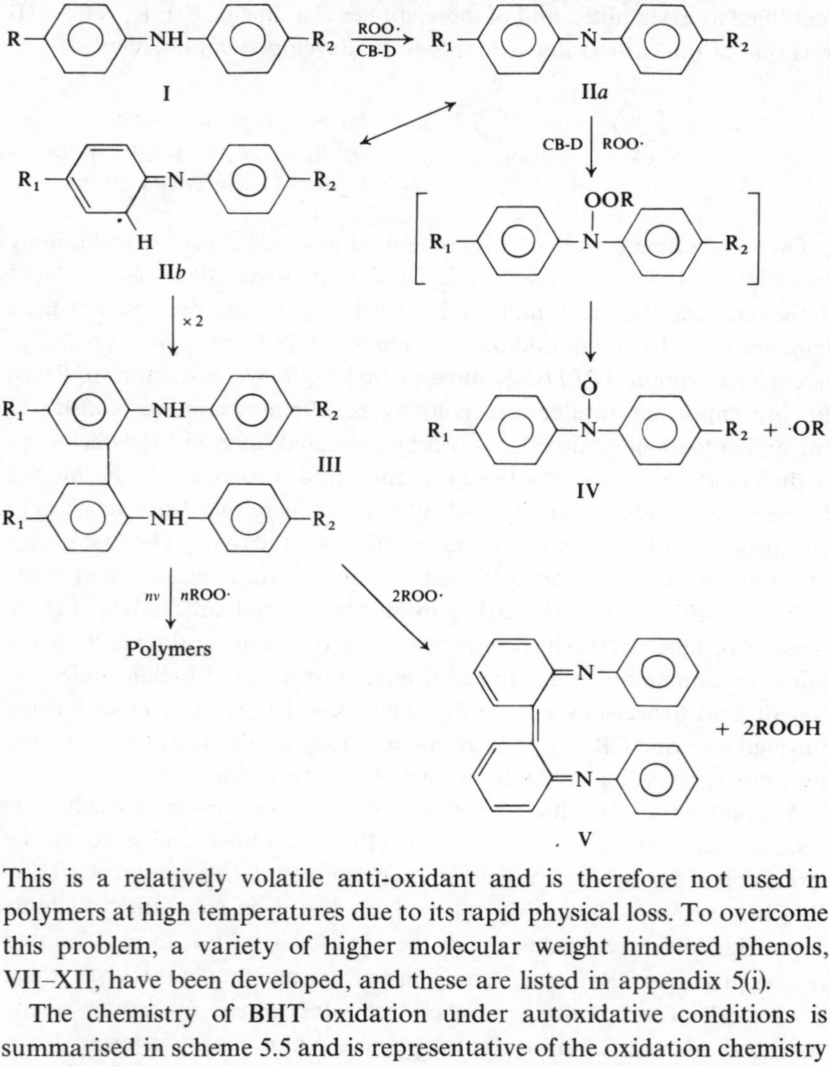

This is a relatively volatile anti-oxidant and is therefore not used in polymers at high temperatures due to its rapid physical loss. To overcome this problem, a variety of higher molecular weight hindered phenols, VII–XII, have been developed, and these are listed in appendix 5(i).

The chemistry of BHT oxidation under autoxidative conditions is summarised in scheme 5.5 and is representative of the oxidation chemistry of all hindered phenols. The stoichiometric inhibition coefficient, f^*, for hindered phenols is normally 2. However, the peroxydienone, XV, formed by trapping of a second alkylperoxyl radical is thermally and photolytically unstable, and leads to the formation of radicals which can re-initiate the process. The quinonoid compounds XVII and XX are formed as major by-products from the initially formed aryloxy radical (XIV).

* f is defined as the number of chains terminated per mole of mono-functional anti-oxidant.

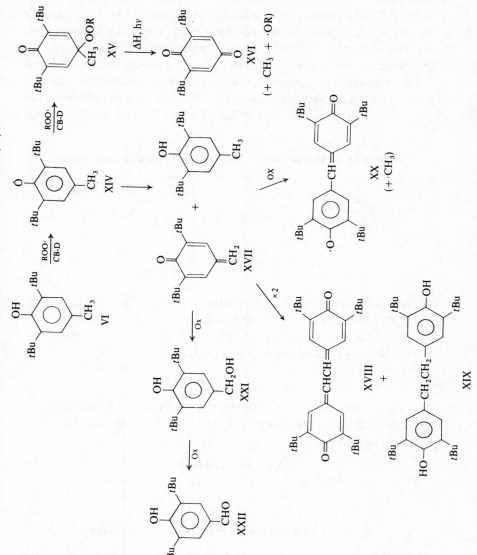

Scheme 5.5 Oxidative transformations of BHT (VI).

The activities of phenolic anti-oxidants are strongly dependent on their structures. In the general case the transition state involves both electron delocalisation and charge separation (scheme 5.6). Consequently, groups in the 2, 4 and 6 positions which extend the delocalisation of the unpaired electron (e.g. phenyl or methyl) increase activity. Electron-releasing groups decrease the energy of the transition state and consequently increase antioxidant activity, whereas electron attracting groups decrease activity.

The presence of at least one tertiary alkyl group in the ortho position is necessary for high antioxidant activity. Many of the most effective antioxidants are substitued in both ortho positions by tertiary alkyl groups.

Scheme 5.6 Transition state in the oxidation of phenolic antioxidants.

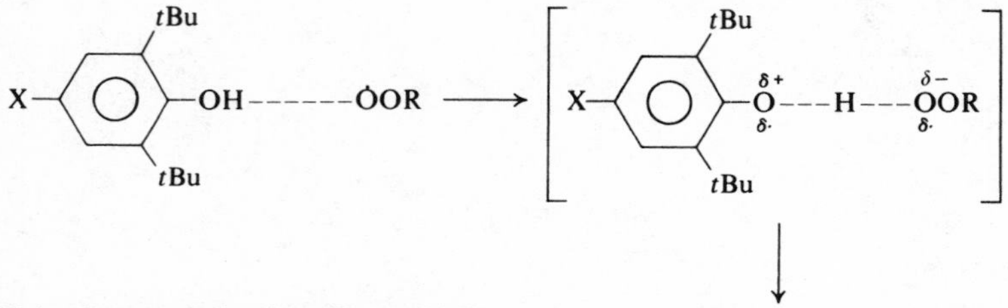

When X is electron releasing (R_2N, RNH, RO, R, etc.) antioxidant activity is increased.

When X is electron attracting (Cl, CN, COOH, NO_2, etc.) antioxidant activity is decreased.

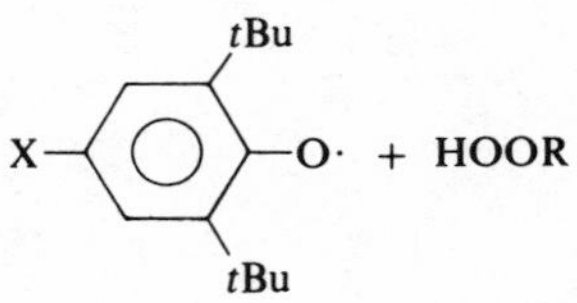

This steric enhancement of antioxidant activity is due to the increased stability of the derived phenoxyl radical which reduces the rate of the chain transfer reaction, (5.1)

$$A \cdot + RH \rightarrow AH + R \cdot \tag{5.1}$$

5.2.2. Chain-breaking acceptor (CB-A) antioxidants

Antioxidants which act by oxidising or trapping alkyl radicals are generally inhibitors for polymerisation in the absence of oxygen. They include benzoquinone and aromatic nitro compounds, but by far the most important are the 'stable' free radicals, of which galvinoxyl (XXIII) and nitroxyls (e.g. IV and XXIV) are the most effective.

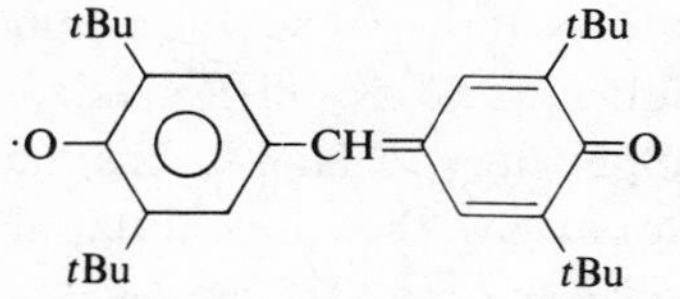

XXIII,G 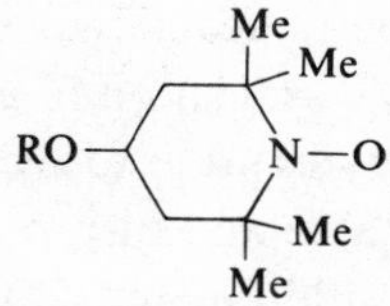   XXIV

It will be seen later that 'stable' aryloxyl and nitroxyl radicals behave as catalytic antioxidants by the cyclical process outlined in scheme 5.3.

5.2.3. Peroxidolytic antioxidants

A variety of compounds catalyse the decomposition of hydroperoxides but many of these involve, to a greater or lesser extent, the formation of free radicals. Most of these, notably transition metal salts, are therefore pro-oxidants rather than antioxidants (see chapter 4).

High molecular weight phosphite esters have been used for many years as 'gel stabilisers' for synthetic rubbers during storage. Tris-nonyl phenyl phosphite (XXV), a typical commercial antioxidant,

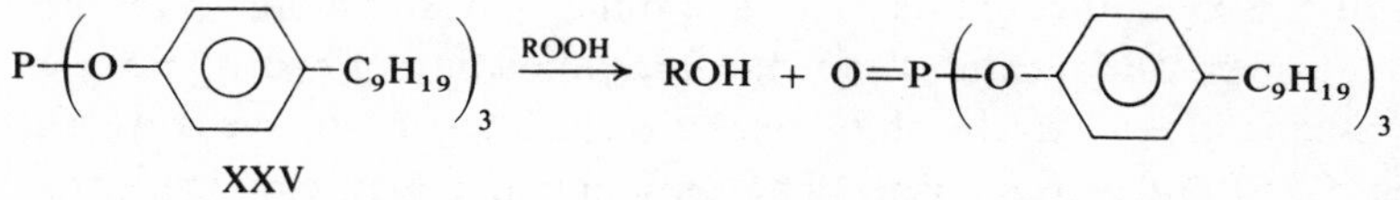

XXV

$$\tag{5.2}$$

stoichiometrically reduces hydroperoxides to alcohols (PD-S). Of much more general applicability is the class of sulphur compounds which destroy hydroperoxides by a catalytic mechanism (PD-C). Acidic species (e.g. sulphur acids) are very effective antioxidants in model substrates but cannot normally be used in polymers due to their insolubility in organic media and their corrosive tendencies. However, a number of sulphur compounds give acidic species on oxidation with hydroperoxides and therefore act as reservoirs for the peroxidolytic species.

Of particular importance because they are the basis of most thermal anti-oxidant systems for thermoplastic hydrocarbon polymers, are the dialkyl thiodipropionates, XXVI.

$$ROCOCH_2CH_2SCH_2CH_2COOR$$

XXVI

(a) $R = C_{12}H_{25}$, DLTP
(b) $R = C_{18}H_{37}$, DSTP

The most important features of their complex oxidation chemistry are summarised in scheme 5.7. The oxidation of XXVI to sulphoxide XXVII is a stoichiometric reaction. At temperatures above 60 °C, XXVII undergoes a reversible dissociation to give a sulphenic acid, XXVIII, which is unstable and eliminates water to give a thiosulphinate ester XXIX. This compound is itself unstable and disproportionates to give the stable disulphide (XXX) and thiosulphonate (XXXI). However, in the presence of excess hydroperoxide, the sulphenic acid XXVIII is oxidised to the sulphinic acid (XXXII) and further to the sulphonic acid (XXXIII), both of which can be identified as reaction products. All the sulphur acids (XXVIII, XXXII and XXXIII) have two modes of antioxidant action. They all readily inhibit oxidation initiated by an alkyl radical generator, azobis-isobutyronitrile, AIBN (XXXIV) which gives rise to alkylperoxyl radicals in the presence of oxygen.

$$CH_3-\underset{\underset{CN}{|}}{\overset{\overset{CH_3}{|}}{C}}-N{=}N-\underset{\underset{CN}{|}}{\overset{\overset{CH_3}{|}}{C}}-CH_3$$

XXXIV

However, compared with arylamines or hindered phenols they are not powerful CB-D anti-oxidants. The sulphinic and sulphonic acids are however also powerful catalysts for the decomposition of hydroperoxides and consequently effectively inhibit oxidation by interrupting the B cycle in scheme 5.1 (PD-C mechanism). The antioxidant activity of methyl-β-sulphinopropionate, MSP (XXXII, R = CH$_3$) in a hydroperoxide-initiated autoxidation is illustrated in Fig. 5.1. This shows that pro-oxidant and anti-oxidant effects are in competition, depending on the conditions. This is a

Figure 5.1. Oxidation of cumene initiated by cumene hydroperoxide (0.1 M) in the presence of methyl-β-sulphinopropionate (MSP) and its pyridine salt. (Reproduced by kind permission of *Europ. Polym. J.*, **15**, p. 241, 1979.)

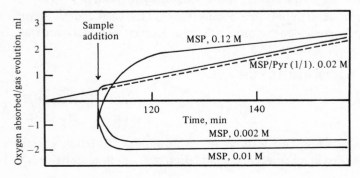

Scheme 5.7 Formation of peroxidolytic antioxidants by the oxidation of thiodipropionate esters.

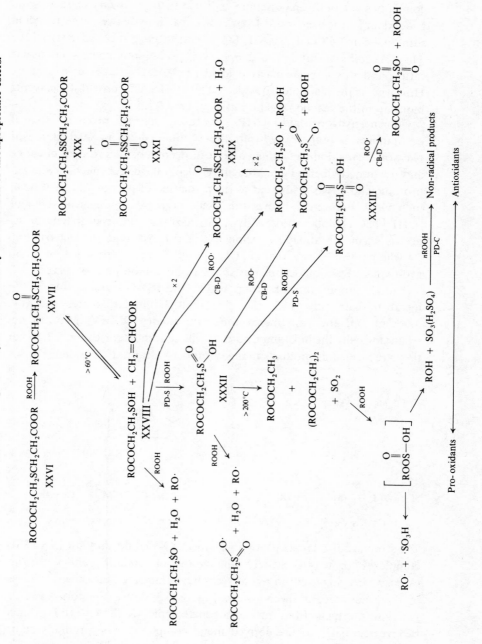

characteristic feature of sulphur-based peroxidolytic antioxidants and constitutes one of the major complexities in their practical use. In general, pro-oxidant processes result from redox reactions between the reducing sulphur acids (XXVIII, XXXII, SO_2, H_2SO_3, etc.) and hydroperoxides. They predominate at low hydroperoxide to sulphur compound molar ratios, and this is particularly evident at the early stages of oxidation. However, as the reaction proceeds, the more effective peroxidolytic agents build up in the reaction medium (see scheme 5.7.).

Cumene hydroperoxide (CHP) gives quite different products when it decomposes homolytically from those obtained in an acid catalysed reaction. Transition metal ions catalyse its decomposition to acetophenone and α,α-dimethylbenzyl alcohol (cumyl alcohol) (see scheme 5.8) whereas protonic and Lewis acids lead to the formation of phenol and acetone in high yield. Figure 5.2 shows the yields of these products at different $[CHP]/[SO_2]$ molar ratios. A sharp change from homolysis to heterolysis occurs at a molar ratio in the region of 1. The same change is reflected in Fig. 5.1 where it can be seen that at $[CHP]/[MSP]$ molar ratios <1 an initial pro-oxidant reaction occurs, but above 1 only inhibition is observed.

Another important sub-group within the PD-C classification is the metal thiolates of which the zinc dialkyldithiocarbamates (XXXIV, $M = Zn$) and the zinc mercaptobenzthiazolates (XXXV, $M = Zn$) are technologically the most important. It will be seen later (section 5.3.2) that they are formed as products in many rubber vulcanisation systems.

XXXIV   XXXV

XXXVI   XXXVII

The structurally related dithiophosphate (XXXVI) and xanthate (XXXVII) complexes show an entirely analogous antioxidant behaviour and extensive investigations in recent years have elucidated the salient features of their antioxidant mechanisms. Figure 5.3 shows the formation of reaction products from mercaptobenzthiazole (XXXIX, MBT) which behaves similarly to the derived metal complexes. The products of its reaction with hydroperoxides vary with the molar ratio $[ROOH]/[MBT]$.

Scheme 5.8 Alternative decomposition products formed in the homolytic and ionic decomposition of CHP.

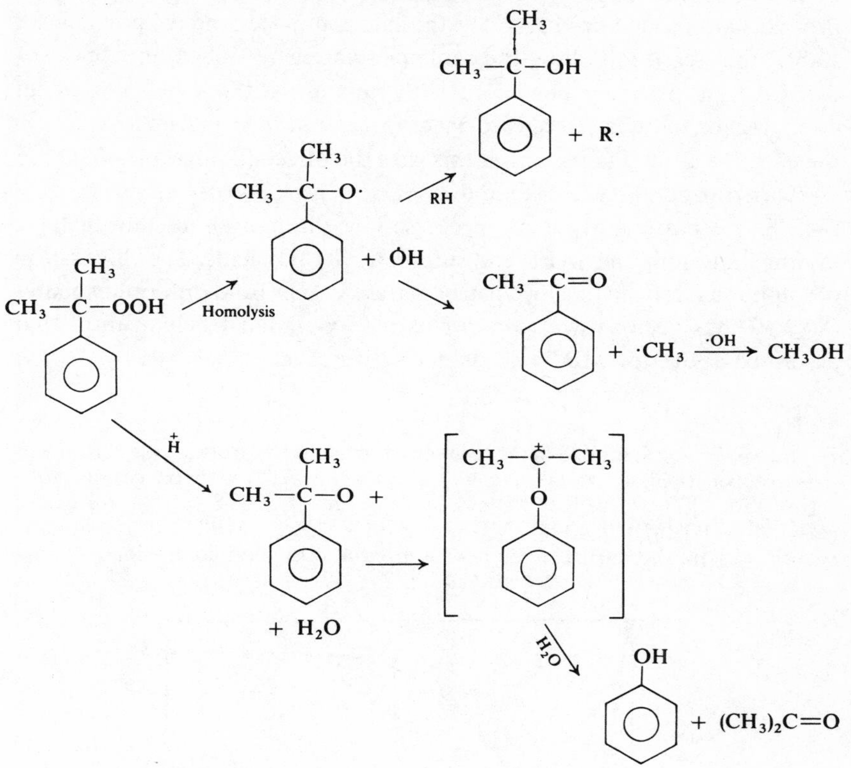

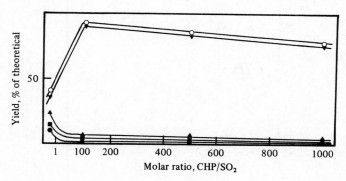

Figure 5.2. Effect of cumene hydroperoxide/sulphur dioxide molar ratio on the product distribution in the reaction of cumene hydroperoxide with sulphur dioxide. ○ phenol; ▲ acetone; ▼ α-methyl styrene; ■ α,α-dimethyl benzyl alcohol; ● acetophenone. (Reproduced by kind permission of *Europ. Polym. J.*, **15**, p. 249, 1979.)

At low ratios, the disulphide, MBTS (scheme 5.9) is the major product, but in the presence of excess hydroperoxide, sulphur acids and their breakdown products are the end products. The thiolate complexes and (where they are stable) the free thiols show the same characteristic transition from pro-oxidant to antioxidant behaviour with time as do the thiodipropionate esters. Again, the acidic species formed by reaction with hydroperoxides are the effective antioxidants, but in this case the direct elimination of SO_2 to give benzthiazole (XLII, BT) and of SO_3 to give 2-hydroxybenzthiazole (XLIV) are particularly facile processes by thermolysis of the unstable intermediate sulphinic (XLI) and sulphonic (XLIII) acids. The elimination of sulphur acids from the dithiocarbamates (XXXIV) and dithiophosphates (XXXVI) has been shown to occur by a very similar mechanism to that described above for MBT and its derivatives.

Figure 5.3. Products formed in the reaction of *tert*-butyl hydroperoxide (TBH) with mercaptobenzthiazole (MBT) and its disulphide (MBTS) with increasing molar ratio [TBH]/[S], where S is the sulphur species. (*a*) MBTS (MBT); (*b*) MBTS (MBTS); (*c*) BT (MBT); (*d*) BT (MBTS); (*e*) BTSO (MBT, MBTS). Compounds in parentheses are the starting materials. (Reproduced by kind permission of *Europ. Polym. J.*, **15**, p. 241, 1979.)

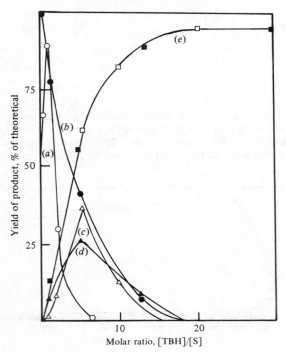

Scheme 5.9 Transformation of mercaptobenzthiazole to acidic species by reaction with hydroperoxides.

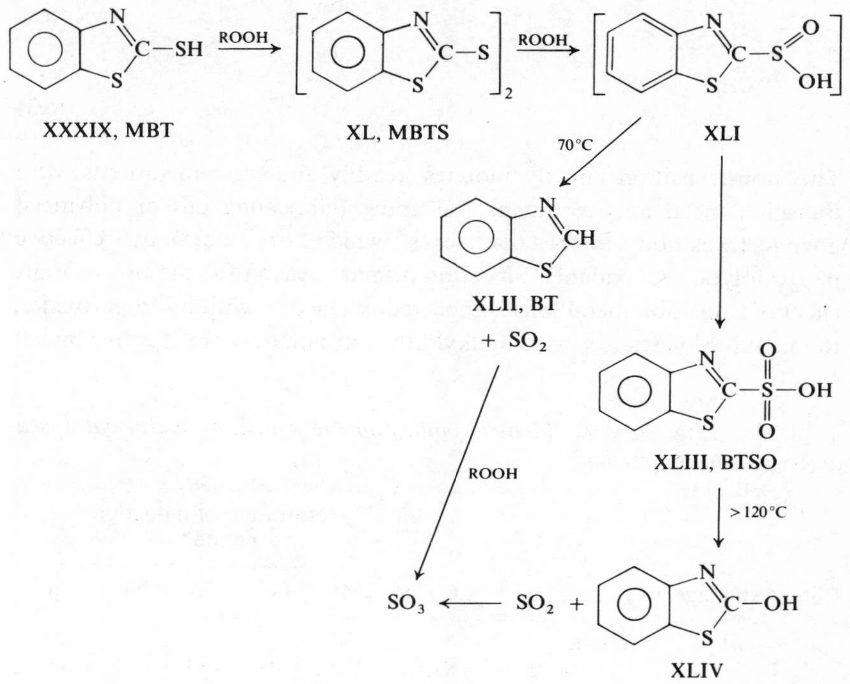

XXXIX, MBT   XL, MBTS   XLI

XLII, BT
+ SO$_2$

XLIII, BTSO

SO$_3$ ⟵ SO$_2$ + XLIV

5.2.4. Metal deactivators

A variety of chelating agents have been found to be effective inhibitors for autoxidation catalysed by transition metal ions. The most effective in hydrocarbons are the Schiffs bases derived from salicaldehyde. Tetradentate disalicydieneethylenediamine (XLV) has been found to be an effective deactivator for Cu (table 5.1), but it is an activator for metals with a higher co-ordination number (e.g. Fe, Co and Mn). An octadentate ligand (XLVI) with the same basic structure is able to deactivate completely all the common transition metal ions (table 5.1), and other octadentate ligands have been found to behave in the same way.

The effect of a metal complex is rarely as simple as that described above. Very frequently the ligand itself has a superimposed antioxidant function. Thus, for example, the nickel acetophenone oxime complex (XLVII) has the ability to destroy hydroperoxides stoichiometrically, reaction 5.3, a reaction which occurs particularly readily in light. It is, therefore, an effective UV stabiliser.

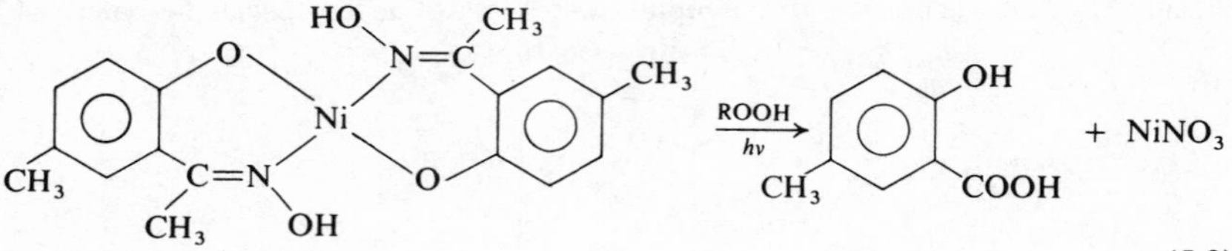

$$\text{XLVII} \xrightarrow[hv]{ROOH} \quad + NiNO_3 \tag{5.3}$$

The non-transition metal thiolates readily undergo metathesis with transition metal ions in organic substrates (lubricating oils or polymers) giving transition metal complexes which are normally effective peroxidolytic antioxidants. Since the primary reason for the pro-oxidant effect of transition metal ions is their redox reaction with hydroperoxides, the metal thiolates (e.g. zinc dialkyldithiocarbamates) are effective 'metal

Table 5.1. *Effect of salicaldehyde imines on the transition metal catalysed autoxidation of petrol.*

Complexing agent	Code	% restoration of induction period*				
		Mn	Fe	Co	Ni	Cu
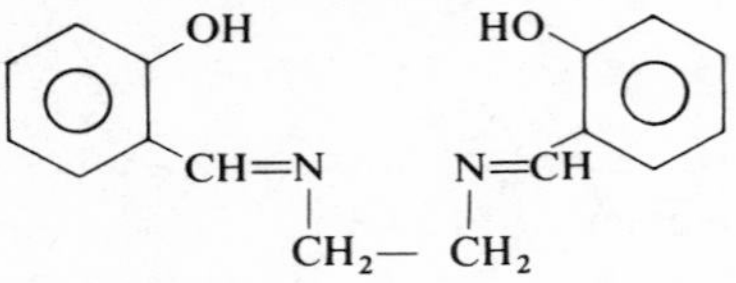	XLV	−103	−43	−883	—	+100
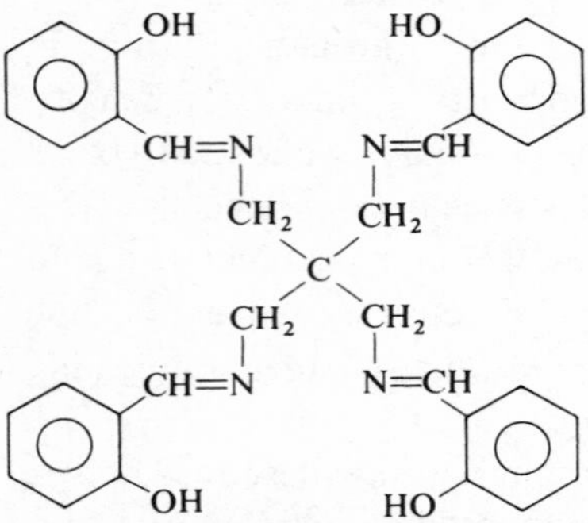	XLVI	+100	+100	+100	+100	+100

* A positive value means that the complexing agent extends the induction period relative to that found for the transition metal ion alone. +100 means 100% restoration of the induction period relative to a control with no transition metal ion. A negative value means that the ligand activates the pro-oxidant effect of the metal ion.

ion inhibitors' primarily because they remove hydroperoxides in an ionic reaction.

5.2.5. *UV screens and filters*

An effective way of protecting polymers from the effects of UV irradiation is to coat the surface with a non-transmitting film of metal or other opaque material. However, one of the major advantages of plastics is that they do not corrode and therefore do not normally require painting. Moreover it is not always possible to post-treat a polymer artifact (e.g. a film) in this way. A good deal of research has therefore been devoted to the development of UV absorbing additives which can be added to the polymer during normal fabrication procedures. Carbon black is singularly successful in this respect. Not only does it absorb all incident light but it appears to have a positive additional antioxidant effect due to the presence of phenolic and quinonoid groups in its structure. In spite of its success, the use of carbon black is limited by its colour. Other light-screening pigments such as titania are widely used in packaging but are not as effective as carbon black as light screens, and frequently a mixture of titania and carbon black (which gives a grey appearance) is used in outdoor fittings (PVC drainpipes, gutters, etc.).

In recent years considerable research in industrial laboratories has led to the discovery of non-coloured UV filters. These are essentially UV-absorbing 'dyestuffs' which are consequently not coloured in visible light. Two main chemical structures XLVIII and XLIX have been developed as commercial UV absorbers for plastics.

XLVIII   XLIX

2-hydroxy-4-octoxybenzophenone (XLVIII) is interesting from the chemical point of view since the analogous benzophenones without a 2-hydroxy group are effective sensitisers for photo-oxidation (chapter 4, section 4.2.3). Both classes exhibit a particularly strong absorption in the region of 330 nm and this is found to be associated with the hydrogen bonded interaction between the phenolic hydroxyl group and adjacent double-bonded groups ($C=O$ and $C=N$). Alkylation of the free hydroxyl group reduces the absorption very substantially and these derivatives are not UV stabilisers.

Unlike the unsubstituted benzophenone or benzotriazole, the UV absorbers do not give rise to a triplet state on absorbing light and are consequently much more UV-stable. It appears then that the light is absorbed by the ketone group in the normal way but the triplet state is internally deactivated by hydrogen transfer in rather the same way as an alkoxyl or peroxyl radical hydrogen abstracts from a phenol (see scheme 5.10), and indeed there is evidence that the 2-hydroxybenzophenones are weak CB-D antioxidants. The quinonoid structure, XLVIII(*b*), undergoes tautomerism to the more stable hydroxybenzophenone; the UV energy initially absorbed is thus emitted as heat.

Some transition metal thiolates (e.g. XXXIV, XXXVI and XXXVII, where M = Ni or Co) are also able to dissipate absorbed UV radiation and emit it harmlessly as thermal energy. However, the evidence suggests that UV screening plays only a minor part in their UV-stabilising function which is primarily associated with their peroxidolytic (PD-C) activity. However, to be effective antioxidants under conditions of UV irradiation necessitates that UV stabilisers must be stable to UV light, and it is this characteristic which distinguishes some metal complexes, notably nickel chelates, from others (e.g. Zn or Fe) which are effective thermal anti-oxidants but are not UV stabilisers. Another nickel complex which appears to function primarily by an antioxidant mechanism is the nickel *bis*-phenolate, L.

Scheme 5.10 Mechanism of light absorption by 2-hydroxybenzophenone UV stabilisers.

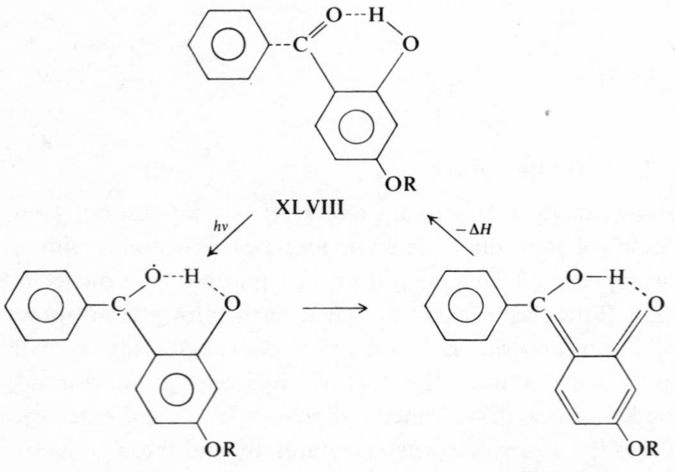

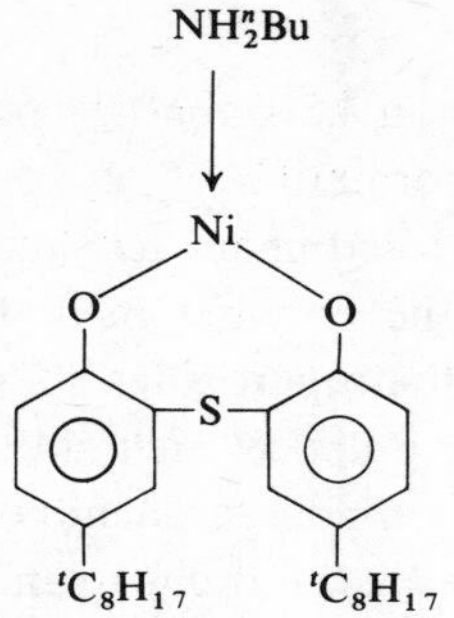

$$NH_2^nBu$$

L

This compound, like the nickel thiolate but unlike the 2-hydroxybenzo-phenones, is a weak thermal antioxidant. Again, like the nickel dithiolates, it is UV-stable and its ability to persist in a polymer under conditions of UV irradiation differentiates it from many other phenolic antioxidants which are not UV stabilisers.

5.3 Stabilisation of polymers during manufacture and in service

Scheme 5.1 indicates that the autoxidation chain reaction can be interrupted in a number of different and complementary ways. All mechanisms do not operate under all conditions in any one polymer, but some antioxidants do act by more than one mechanism and under more than one set of conditions. The object of stabilisation technology is to achieve stability under a wide range of conditions at the lowest possible cost, and most commercial stabilising systems have resulted from an empirical approach to the problem. The object of the antioxidant scientist is to understand the mechanisms by which antioxidants act in order to prepare the way for major new advances that are not normally possible on the basis of a purely empirical approach.

5.3.1. Melt stabilisation

Stabilisation of polymers during processing is not only important to the polymer convertor, but it is basic to the subsequent service performance of the polymer. The primary objective is to maintain constant the viscosity of the polymer when it is in the barrel of a screw extruder or in the melt pool during 'spinning' of a fibre. An equally important function of a melt stabiliser is to inhibit the formation of hydroperoxides and other sensitising groups which would otherwise impair the durability of the polymer in service.

Polyolefins

It was seen in chapter 4, section 4.2.1, that shearing of the polymer chain, followed by the subsequent formation of peroxides which initiate radical processes (cross-linking in PE and chain scission in PP), are the primary causes of MFI changes in the polymer melt. Both removal of macro-radicals and destruction of hydroperoxides should therefore be effective melt-stabilising processes. In practice, melt stabilisation of polyolefins is fairly readily achieved using a hindered phenol (CB-D) antioxidant. BHT (VI), most commonly used because it is cheap and relatively efficient, gives

Figure 5.4. Variation of galvinoxyl [G·] and hydrogalvinoxyl [GH] concentrations in polypropylene during the induction period to MFI change at 200 °C. 1 [G·]; 2 [GH]; 3 [G·]+[GH]; 4 MFI.

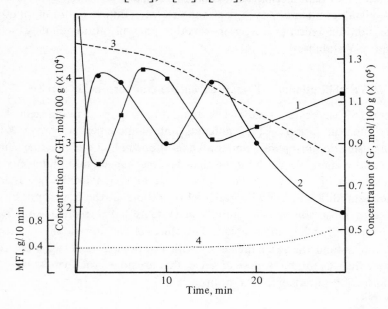

Scheme 5.11 **Oxidation of macro-alkyl radicals in polypropylene by galvinoxyl (G·) during processing.**

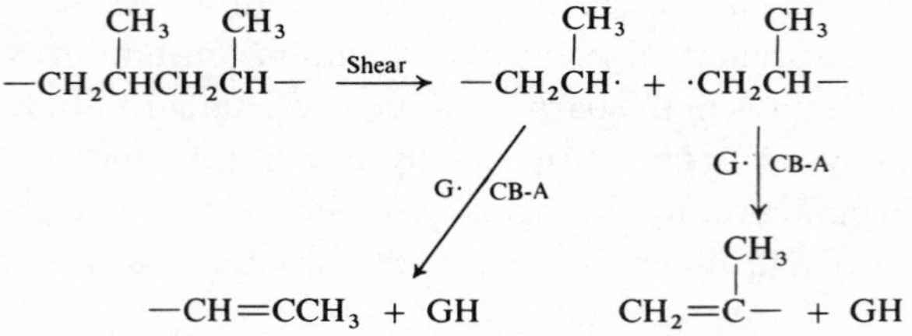

rise to oxidation transformation products (notably XVIII and XX, scheme 5.5) that are much more effective than BHT itself on a molar basis (see table 5.2). Galvinoxyl, G·(XX) is particularly effective in polypropylene and is substantially converted to the parent phenol, hydrogalvinoxyl, GH (LI) during the first minute of mixing in a closed internal mixer at 200 °C (see figure 5.4). The formation of this reduced product results from the oxidation

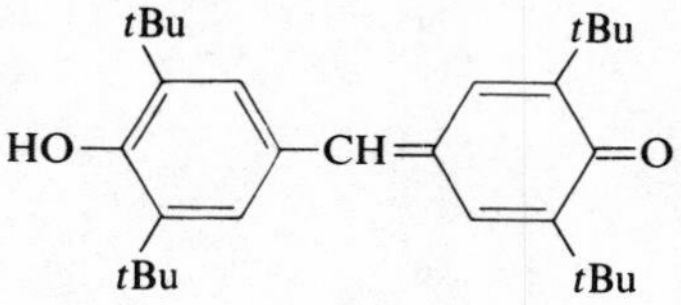

$$HO-\underset{tBu}{\overset{tBu}{\bigcirc}}-CH=\underset{tBu}{\overset{tBu}{\bigcirc}}=O$$

LI, GH

of a macro-alkyl radical by galvinoxyl in a CB-A process (see scheme 5.11). However, as the shear in the mixer decreases due to reduction in the polymer melt viscosity, GH is partially reoxidised to G·(CB-D) by alkylperoxyl radicals, which then accumulate in the system. Both CB-A and CB-D mechanisms occur together, giving rise to a regenerative cycle (scheme 5.12) analogous to that typified generally in scheme 5.3. Each cycle is accompanied by the formation of one double bond in the polymer chain, (scheme 5.12), and by measuring the amount of unsaturation formed during the induction period to MFI change per mole of antioxidant, the stoichiometric inhibition coefficient, f, is found to be about 50. Hindered nitroxyl radicals (e.g. XXIV), iodine and cupric stearate behave in the same way (see table 5.3).

The reduced forms of the antioxidant redox couples in each case can additionally react with hydroperoxides formed in the polymer by scheme

Scheme 5.12 Cyclical regeneration of galvinoxyl (G ·, XXIII) during the melt stabilisation of polypropylene.

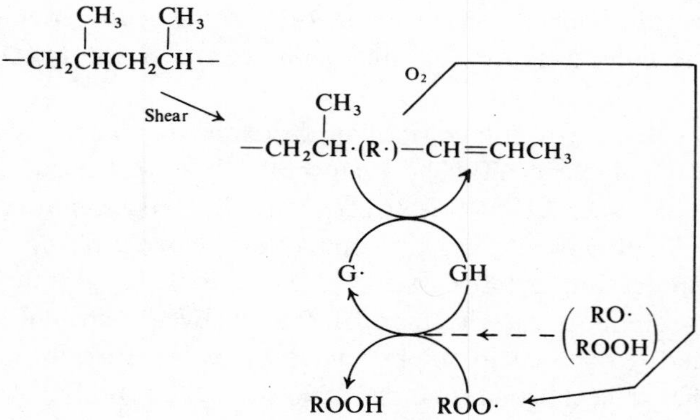

Table 5.2. *Effectiveness of BHT oxidation products as melt stabilisers in polypropylene at 270°C.*

Compound	Code	ΔMFI[+] 4.5*	0.5*
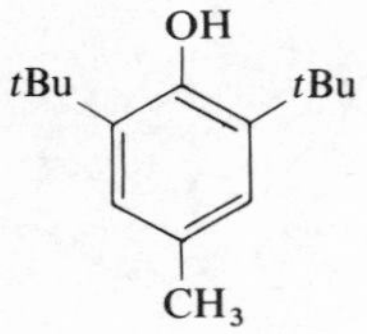 (structure VI: 2,6-di-tBu-4-methylphenol, OH, tBu, tBu, CH₃)	VI	100	270
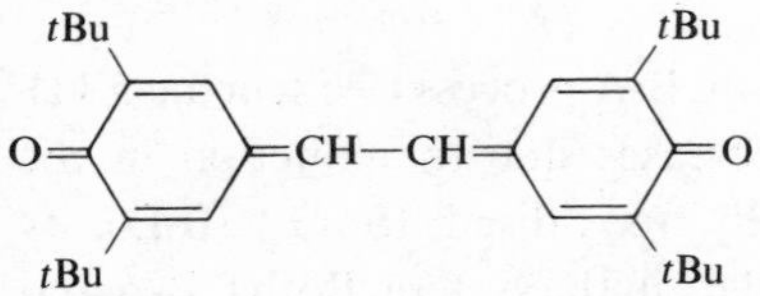 (structure XVIII: tBu, tBu, O=...=CH—CH=...=O, tBu, tBu)	XVIII	50	60
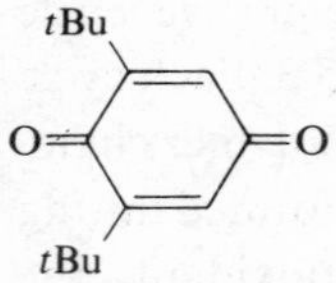 (structure LII: tBu, O=...=O, tBu)	LII	55	65
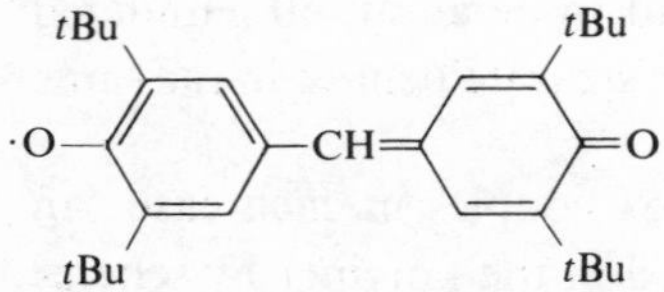 (structure XX: tBu, tBu, ·O—...—CH=...=O, tBu, tBu)	XX	45	45

[+] Increase in MFI, % in a screw extruder.
* Additive concentration, 10^3 mol kg^{-1}.

5.12 with regeneration of the oxidised form of the redox couple. This reaction may contribute to the antioxidant activity of the CB-A/CB-D system.

As anticipated, peroxidolytic (PD) antioxidants are also effective melt stabilisers for polyolefins. Table 5.4 shows that zinc and nickel dinonyl dithiocarbamates (XXXIV, R = iso C_9H_{19}, M = Zn, Ni) are more effective than typical high molecular weight commercial hindered phenols during continuous processing at 190°C. Table 5.5 indicates that in general the PD antioxidants are more effective than BHT itself. Even elemental sulphur which is readily oxidised to sulphur acids by hydroperoxides is highly effective. The cyclic phosphite and phosphate esters (LVII–LIX) have been

Table 5.3. *Effectiveness of redox antioxidants (CB-A/CB-D) as melt stabilisers in polypropylene at 270°C.*

Compound	Code	Redox couple	ΔMFI[+] 4.5*	0.5*
BHT	VI	—	100	270
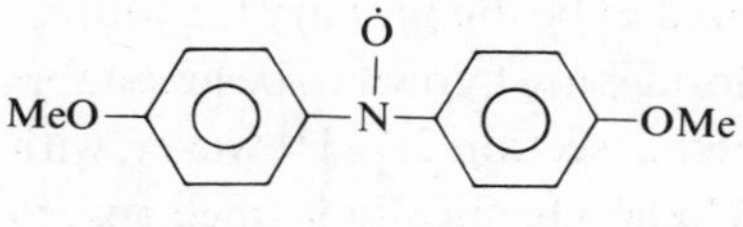 MeO–⟨O⟩–N(=O)–⟨O⟩–OMe	LIII	>NO·/>NOH	—	45
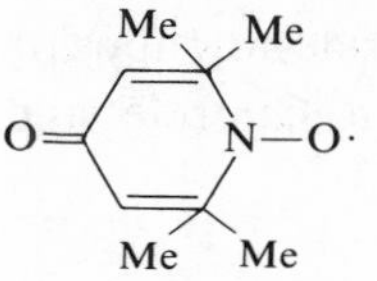 O=[2,2,6,6-tetramethyl]N–O·	LIV	>NO·/>NOH	—	55
I_2	LV	I·/IH	45	50
$Cu(OCOC_{18}H_{37})_2$	LVI	Cu^{2+}/Cu^+	40	80

[+] ΔMFI, % change in melt flow index in a screw extruder.
* Concentration, 10^3 mol kg^{-1}.

Table 5.4. *Comparison of dithiocarbamate complexes with hindered phenols as melt stabilisers for PP (torque rheometer at 190°C; concentration 0.2%[21]; initial MFI, 0.40).*

	Melt flow index at time indicated				
Processing time, minutes:	5	10	20	25	30
Anti-oxidant					
None	0.60	1.20	1.80	2.30	3.10
ZnDNC (XXXIV, R=C_9H_{19})	0.40	0.40	0.40	0.40	0.40
NiDNC (XXXIV, R=C_9H_{19})	0.40	0.40	0.40	0.49	0.54
Irganox 1010 (IX)	0.40	0.40	0.42	0.46	0.51
Irganox 1076 (VIII)	0.40	0.43	0.49	0.52	0.59

shown to behave, like the sulphur compounds, as catalysts for peroxide decomposition.

(b) Polyvinylchloride

It was seen in chapter 4, section 4.2.1 that the thermal degradation of PVC differs from that of the hydrocarbon polymers in that hydrogen chloride eliminated from the polymer forms a redox system with hydroperoxides, thus accelerating the 'unzipping' process which leads to colour in the polymer. A major objective of PVC stabilisation technology must therefore be to reduce the formation of HCl and hydroperoxides in the polymer, as well as to remove the developing unsaturation which is both the cause of colour and the source of further instability. Until comparatively recently, the removal of allylic chlorine was considered to be the primary function of an effective PVC stabiliser, since allylic chlorides are known to be less stable to pyrolysis than alkyl chlorides (see chapter 2, section 2.4). However, with the recognition of the importance of mechano-chemically formed macro alkyl radicals during processing the removal of allylic chloride now appears to be less important than the removal of hydrogen chloride and hydroperoxides.

Table 5.5. *Peroxidolytic (PD) antioxidants as melt stabilisers for PP at 270 °C.*

Antioxidant	Code	ΔMFI 4.5*	0.5*
BHT	VI	100	270
$(C_9H_{19}\text{–}C_6H_4\text{–}O)_3P$	XXV	40	150
(benzo-1,3,2-dioxaphosphole)–P–O–(2,6-di-tBu-4-Me-phenyl)	LVII	60	85
(benzo-1,3,2-dioxaphosphole) P=O with OH	LVIII	75	150
(structure with OH, OH, P–OH, O, O)	LIX	90	—
$(C_{12}H_{25})_2SO$	LX	65	50
$(C_{12}H_{25}OCOCH_2CH_2)_2SO$	LXI	50	110
$(S)_8$	LXII	60	—

* Concentration, 10^3 mol kg^{-1}.

Melt stabilisers for PVC can therefore be classified in the following way:

(*i*) Reagents for hydrogen chloride and allylic chlorides.

The most important examples are the metal carboxylates (LXIII), lead carbonate and tetravalent derivatives of tin, of which the dialkyl tin maleates (LXIV) and (LXV), and the dialkyl tin thioglycollates (LXVI), are the most frequently used.

$$M(OCOR)_2$$

LXIII

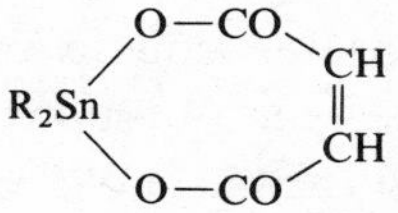

$$R_2Sn \underset{O-CO}{\overset{O-CO}{<}} \underset{CH}{\overset{CH}{>}}$$

LXIV

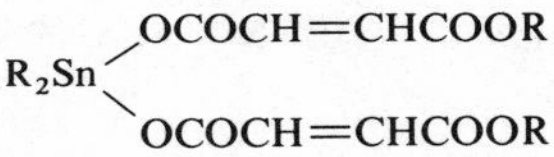

$$R_2Sn \underset{OCOCH=CHCOOR}{\overset{OCOCH=CHCOOR}{<}}$$

LXV

$$R_2Sn \underset{SCH_2COOR}{\overset{SCH_2COOR}{<}}$$

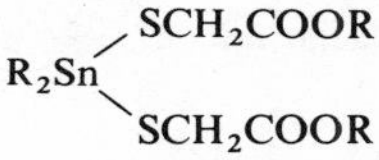

LXVI

The metal soaps (LXIII) are frequently used in synergistic combinations. The reason for this is that, because the zinc and cadmium carboxylates are more electrophilic than the calcium and barium carboxylates, and hence are more readily substituted by Cl, their transformation products ($ZnCl_2$ and $CdCl_2$) are more powerful Lewis acids which actually catalyse the ionic elimination of hydrogen chloride. $CaCl_2$ and $BaCl_2$ on the other hand are relatively inert. The role of calcium and barium carboxylates is therefore to partially regenerate the active electrophyles. This is illustrated for the Cd/Ba combination in scheme 5.13. The use of the two component combination thus extends the useful life of the stabiliser and hence the period to rapid HCl loss.

Scheme 5.13 Mechanism of synergism between cadmium and barium carboxylates in the melt stabilisation of PVC.

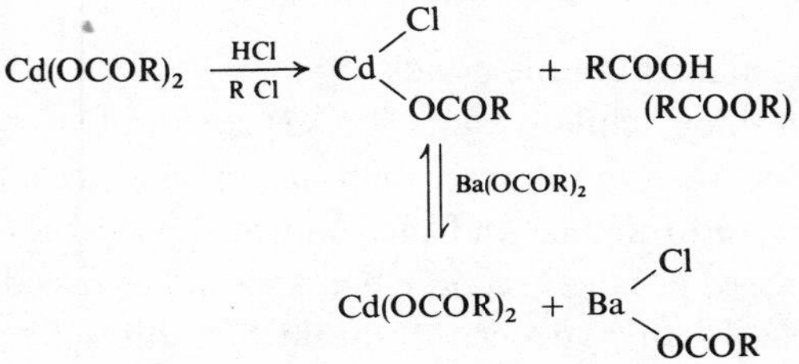

$$Cd(OCOR)_2 \xrightarrow[R\,Cl]{HCl} Cd \underset{OCOR}{\overset{Cl}{<}} + RCOOH$$
$$(RCOOR)$$

$$\Updownarrow Ba(OCOR)_2$$

$$Cd(OCOR)_2 + Ba \underset{OCOR}{\overset{Cl}{<}}$$

Dialkyl tin carboxylates are in general more effective than the metal group II carboxylates as reactive chlorine reagents. The dialkyl tin maleates

(LXIV and LXV) are particularly effective and have in recent years assumed a central role in PVC stabilisation for packaging applications where high clarity and good long-term performance are required.

(*ii*) Diels–Alder reagents for polyconjugated unsaturation.

Although the primary role of the tin maleates is to react with reactive chlorine (either HCl or allylic chlorine), an equally important reaction during processing is the Diels–Alder reaction with the developing conjugated unsaturation (scheme 5.14, reaction (*d*)). This process interrupts the conjugated unsaturation which is the main cause of colour in the polymer. Post treatment of PVC degraded during processing with a diallyl tin maleate leads to almost complete discharge of the colour.

Scheme 5.14 Reactions of the dibutyl tin maleate (DBTM) with PVC during processing.

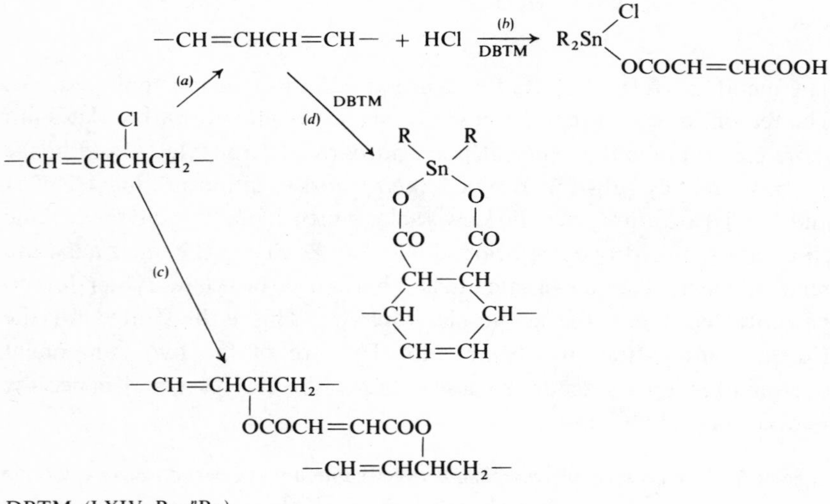

$$-CH=CHCH=CH- \;+\; HCl \xrightarrow[DBTM]{(b)} R_2Sn\!\!\begin{array}{l} Cl \\ OCOCH=CHCOOH \end{array}$$

(a)

(d) DBTM

Cl
|
$-CH=CHCHCH_2-$

R   R
\ /
Sn
/ \
O   O
| |
CO   CO
\   /
CH—CH
$-$CH   CH$-$
\   /
CH=CH

(c)

$-CH=CHCHCH_2-$
|
OCOCH=CHCOO
|
$-CH=CHCHCH_2-$

DBTM, (LXIV, R = ^{n}Bu)

(*iii*) Reagents for isolated double bonds.

The dialkyl tin thioglycollate esters (LXVI) constitute a second class of tin stabilisers, which are more effective than the tin maleates under oxidative conditions. Their performance and mechanism during the service life of PVC will be discussed in a later section, but it should be noted here that not only do they react with hydrogen chloride but thiols liberated in this reaction form adducts with the double bonds which result from the initial mechano-chemical process (see scheme 5.15). The kinetics of peroxide and unsaturation formation are shown in Fig. 5.5 which shows that the octyl tin

Scheme 5.15 Reaction of dioctyl tin thioglycollate (DOTG) with PVC during processing.

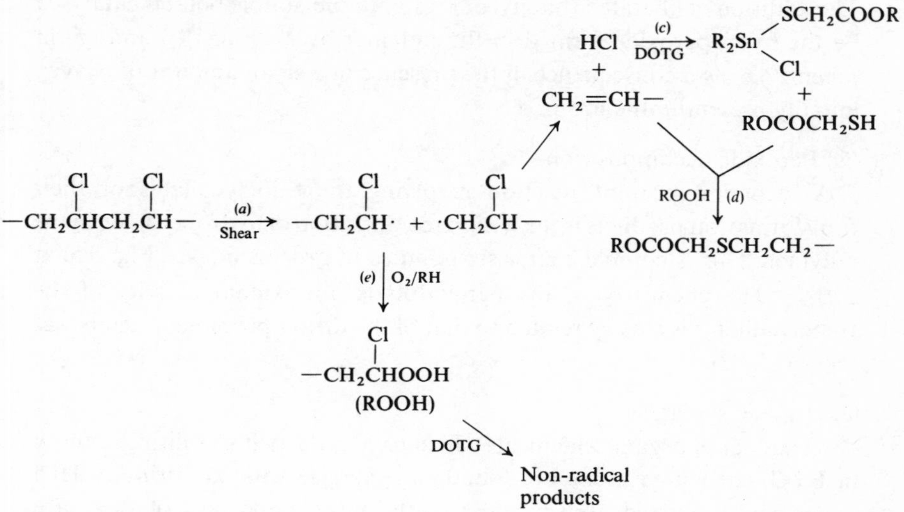

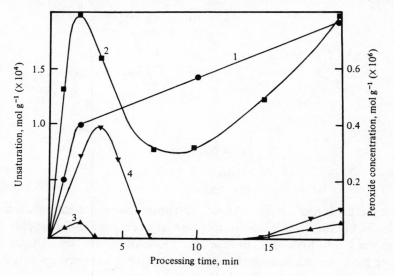

DOTG, (LXVI, R = $^sC_8H_{17}$)

Figure 5.5. Effect of processing at 170 °C on the formation of functional groups in PVC. 1 unsaturation in the absence of additives; 2 peroxides in the absence of additives; 3 unsaturation in the presence of dioctyl tin thioglycollate (DOTG); 4 peroxides in the presence of DOTG. DOTG concentration 1.16×10^{-2} mole/100 g.

thioglycollate (DOTG) does not inhibit the initial formation of these chemical species, but it does lead to their rapid removal during processing. The addition of liberated thioglycollic ester to the double bond is catalysed by the hydroperoxide formed in the polymer by reactions (*e*) and (*f*) in scheme 4.6, as a consequence of the presence of a small amount of oxygen initially present in the mixer.

(*iv*) Peroxide decomposition.

A second important reaction involving the thioglycollates or their transformation products is the ionic decomposition of hydroperoxide in the polymer. This is complete after six minutes of processing (see Fig. 5.5) at 170 °C. The chemistry of the peroxidolytic antioxidant activity of the thioglycollates is closely related to that of the thriodipropionate esters (see section 5.2.3).

(*v*) Organic synergists.

A number of organic chemicals which have little melt stabilising activity in PVC when used alone are found to synergise with the primary HCl scavengers discussed above. Some of the more important of these are epoxides (LXVII), of which epoxidised soya bean oil is an important example, phosphite esters (LXVIII), particularly the mixed dialkyl aryl phosphites, α-phenyl indole (LXIX), diketones (e.g. benzoylacetone (LXX) and β-amino-crotonate esters (LXXI)).

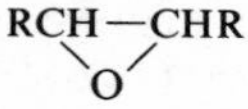

$$RCH\!-\!CHR \qquad\qquad (RO)_2POR'$$
$$\diagdown O \diagup$$

LXVII   LXVIII   R = aryl
                                   R' = alkyl

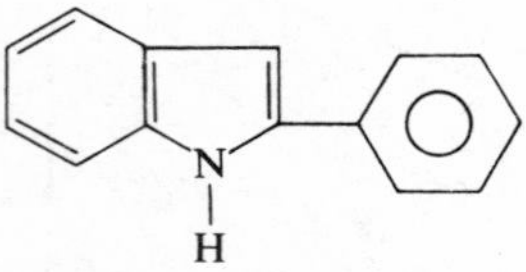

LXIX 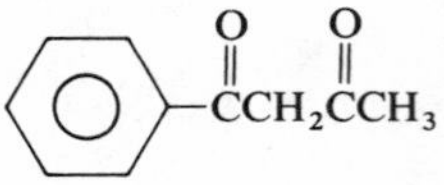   LXX

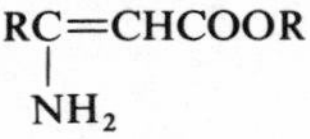

$$RC\!=\!CHCOOR$$
$$|$$
$$NH_2$$

LXXI

The mechanism of action of these compounds has not been completely clarified but they are all able to react with labile chlorine and most of them also react with hydrogen chloride. The phosphite esters certainly act by more than one mechanism. They are effective peroxide decomposers but they also readily add to conjugated carbonyl groups formed during processing to give a phosphonate ester in the presence of HCl (reaction 5.4).

$$-\underset{\underset{O}{\|}}{C}-CH=CH- \xrightarrow{P(OR)_3} \left[-\underset{\underset{\underset{\cdot}{Q}}{|}}{C}=CH-\underset{|}{CH}- \quad + \; P(OR)_3 \right]$$

$$\Big\downarrow HCl$$

$$-\underset{\underset{O}{\|}}{C}-CH_2-\underset{\underset{O=P(OR)_2}{|}}{CH}- \quad [+ \; RCl] \tag{5.4}$$

They thus have the ability to remove one of the more important groups responsible for colour in the polymer.

5.3.2. Thermal oxidative stabilisation

Although polymers are not normally used at temperatures above 100 °C, an 'oven-ageing' accelerated test is almost always used as one of the 'screening' tests that antioxidants have to satisfy. The relevance of this kind of test, which is normally carried out between 70 °C (for unsaturated rubbers) and 140 °C (for saturated polymers such as polyolefins, polyesters, polyamides, etc.), is far from clear since it has been shown that many highly effective antioxidants at moderate temperatures become relatively ineffective at high temperatures, whereas some relatively ineffective antioxidants at moderate temperatures retain their effectiveness much better at high temperatures.

Polyolefins

Table 1.1 (chapter 1) shows the relative effectiveness of a number of different phenolic antioxidants for polypropylene at several temperatures. It is evident that their order of effectiveness changes with temperature and that it is not valid to extrapolate the relative order of effectiveness of antioxidants from high temperature to use temperature. However, it is accepted that from a practical point of view, accelerated tests are necessary since it is not normally possible, for reasons of time, to age polymer products to destruction in their use environment before putting them into service. The present range of commercial antioxidants for polyolefins was developed on the basis of accelerated screening tests. The justification for this is that long-term tests under use conditions, particularly in rubber products, have generally confirmed their effectiveness determined by accelerated tests. It seems likely therefore, that an oven-ageing test will remain one of the hurdles that an antioxidant will have to pass before selection for service trials.

Chain-breaking donor (CB-D) antioxidants generally form the basis of heat-stabilising systems for polyolefins and since the arylamines are normally too discolouring, the hindered phenols are widely used. It might be expected that hindered phenols containing the same functional group (e.g. the widely used 2,6-ditertbutyl phenol structure) should show similar activity on a molar basis in polymers. However, this is not so. Table 5.6 shows that all members of the homologous series, LXXII, have similar

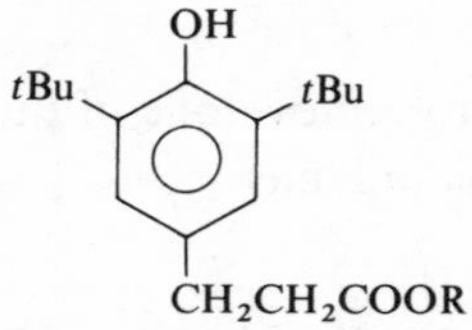

$$tBu \quad \overset{OH}{\underset{CH_2CH_2COOR}{\bigcirc}} \quad tBu$$

LXXII

antioxidant activity at 140 °C in a closed system in a liquid hydrocarbon. However, only the highest molecular weight homologue (LXXII, $R = C_{18}H_{37}$) shows any activity in an air-oven test at the same temperature. The test employed involves the passage of an air stream over the surface of the polymer sample in the form of a thin film. Table 5.6 shows that anti-

Table 5.6. *Effectiveness of antioxidants (LXXII) at the same molar concentration* $(2 \times 10^{-3} \ mol/100 \ g)$ *in decalin and polypropylene at*

LXXII	Mol wt	$T^{1/2}$, h	S, g/100 g	Induction period (hours)		
				D^c	PP^c	PP^o
R						
CH_3	292	0.28	32	25	95	2
C_6H_{13}	362	3.60	∞	23	312	2
$C_{12}H_{25}$	446	83.00	∞	20	420	2
$C_{18}H_{37}$	530	660.00	64	20	200	165
BHT (VI)	220	0.10	100	150	140	2

$T^{1/2}$ Antioxidant half life in PP in N_2 stream at 140 °C.
D^c Induction period in decalin by oxygen absorption at 140 °C.
PP^c Induction period in PP film by oxygen absorption at 140 °C.
PP^o Induction period in PP film in a moving air stream at 140 °C by carbonyl measurement.
S Solubility in hexane at 25 °C.

oxidant effectiveness is dependent upon the ability of the antioxidant to remain in the polymer under these severe conditions.

Consequently all the hindered phenol antioxidants available commercially for polyolefins are large molecules with low volatility (see appendix 5(*i*)). Low molecular mass phenols used in other substrates at lower temperatures (e.g. BHT, VI) are relatively ineffective in polypropylene at 140 °C. The highest molecular mass antioxidant in the series LXXII, $(R = C_{18}H_{37})$ is one of the commercial products, Irganox 1076 (VIII) listed in appendix 5(i). It should be noted however, that in a closed system (oxygen absorption in polypropylene), Irganox 1076 is not the most effective member of the LXXII series; the dodecyl homologue is over twice as effective on a molar basis. This difference almost certainly results from the higher solubility of the dodecyl compound in the hydrocarbon substrate (see table 5.6). In practical terms, results of this kind must be interpreted in the light of the dimensions of the sample being investigated and on the conditions to which it will be subjected in service. In large thick sections, and at relatively low temperatures, the loss of antioxidant should be much less dependent on the volatility of the antioxidant than it is in thin sections (films or fibres) and at high temperatures. For thick samples at ambient temperatures, solubility will be the major physical factor determining antioxidant or stabiliser activity.

It may be concluded then that three factors determine the effectiveness of antioxidants in polymers. These are

(*i*)   intrinsic molar activity;
(*ii*)   substantivity in the polymer;
(*iii*) solubility in the polymer.

A potential means of satisfying the last two criteria is to chemically react the antioxidant with the polymer. By definition, a polymer-bound antioxidant must be molecularly dispersed (i.e. infinitely soluble) and cannot be physically lost from the substrate. This topic will be discussed in a later section.

It was discovered early in the development of polypropylene that sulphur compounds synergise effectively with high molecular mass phenols, giving embrittlement times many times greater than either of the single components alone at the same concentration. This is illustrated for several high molecular mass semi-hindered phenols in combination with dilauryl thiodipropionate (XXVI*a*) in Fig. 5.6. The thiodipropionate esters are not very effective alone since they do not give an induction period when used alone as thermal antioxidants (see Fig. 5.7). It was seen in section 5.2.3 that radical generation occurs in the reaction of DLTP with hydroperoxides before the former is converted to an effective peroxidolytic antioxidant (PD-C). The main role of a hindered phenol, which is the minor component of the system, is to scavenge radicals, thus eliminating this pro-oxidant stage completely.

Figure 5.6. Synergism between dilauryl thiodipropionate (DLTP) and hindered phenols at constant total weight concentration in polypropylene at 150°C. 1 Topanol CA; 2 Tetrakis (4-hydroxy-3-*tert*-2-methylphenyl)butane; 3 Nonox WSP. (Reproduced by kind permission of *Europ. Polym. J.* – Supplement, 1969, p. 189.)

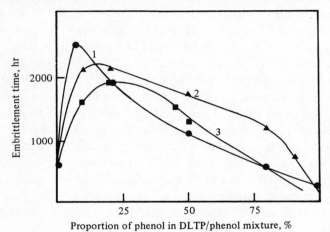

Figure 5.7. Oxidation of polypropylene containing DSTP at 140°C. DSTP concentrations, mole/100 g: 1, none; 2, 2×10^{-4}; 3, 8×10^{-4}; 4, 1.5×10^{-3}; 5, 2×10^{-3}.

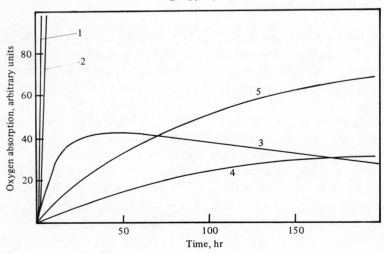

Antioxidants have been designed which contain the types of anti-oxidant activity, (CB-D and PD-C) in the same molecule. LXIII is about twice as effective on a molar basis as LXXII

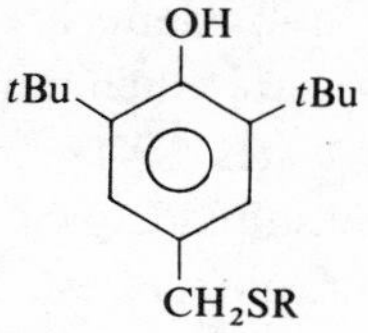

LXXIII   R = alkyl

which does not contain the peroxidolytic sulphide function. Some peroxidolytic antioxidants are much more effective, when used alone in polyolefins, than DLTP. Table 5.7 shows that zinc dinonyldithio-carbamate, XXXIV, $R = {}^iC_9H_{19}$ is a better heat-ageing antioxidant for polypropylene at 140 °C than Irganox 1076 (appendix 5(*i*), VIII). Its effectiveness derives in part from its low volatility, but most importantly, it is much more soluble in the polymer than its ethyl analogue.

It has been shown that antioxidants and stabilisers are excluded from the crystalline regions of the polyolefins and are concentrated in the amorphous regions. It is fortunate therefore that this is the oxidatively-sensitive part of the polymer (see section 4.1.7). A consequence of this is that although unstabilised polyethylene is more resistant to physical degradation than polypropylene, it is relatively easier to stabilise the latter by means of antioxidants.

Table 5.7. *Comparison of dithiocarbamate complexes with a hindered phenol as thermal antioxidants for polypropylene (air oven at 140 °C).*

Antioxidant	Embrittlement time (hours)		
	0.1	0.2	0.4 g/100 g
Irganox 1076 (VIII)	—	60	90
ZnDEC (XXXIV, M = Zn, $R = C_2H_5$)	15	30	80
ZnDNC (XXXIV, M = Zn, $R = {}^iC_9H_{19}$)	55	90	175
NiDEC (XXXIV, M = Ni, $R = C_2H_5$)	12	20	45
NiDNC (XXXIV, M = Ni, $R = {}^iC_9H_{19}$)	20	50	150

Poly(vinylchloride)

PVC is stabilised against the effects of oven ageing by the same agents that are used in processing. However, since thermal oxidation plays a larger part in heat ageing, conventional phenolic antioxidants are frequently used as synergists. The thiotin stabilisers (LXVI) are much more effective thermal antioxidants than are the dialkyl tin maleates (LXIV and LXV). There are two reasons for this. The first is that the sulphide part of the molecule is a highly effective PD-C antioxidant. The acidic species is formed by reaction with hydroperoxides, scheme 5.16, reaction (*a*), (*b*) in a manner analogous to the formation of sulphur acids from DLTP (XXVI(*a*)). The latter compound

Scheme 5.16 Stabilisation mechanism of a dioctyl tin thioglycollate ester (DOTG) during the thermal-oxidative degradation of PVC.

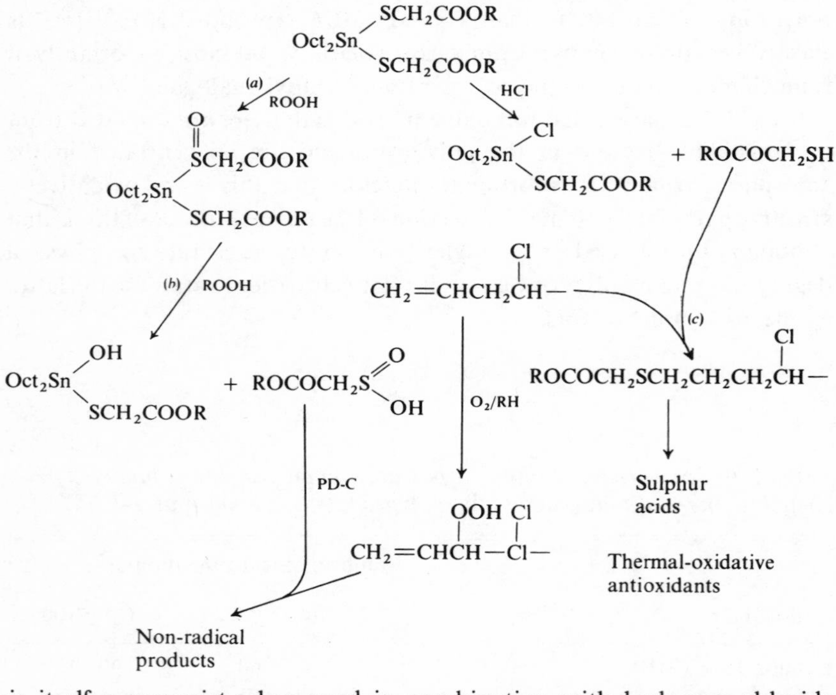

is itself a synergist when used in combination with hydrogen chloride scavengers. The second reason for the effectiveness of the dialkyltin thioglycollates is that the liberated thioglycollic ester adds to developing monoenic unsaturation as it is formed by HCl elimination. This process removes sensitising allylic methylene groups (scheme 5.16, reaction (*c*)). The monosulphides so formed are themselves effective macromolecular antioxidants.

Rubbers

Oxidation-resistant rubbers are frequently produced, not by the use of added antioxidants, but by modifying the vulcanisation formulation. This is normally achieved either by using no elemental sulphur at all (thiuram 'sulphurless' vulcanisate) or by using a much higher ratio of accelerator to sulphur ($\approx 5 : 1$) than would normally be used ($\approx 1 : 5$). The chemical transformation products produced under these conditions are powerful antioxidants (see chapter 4, section 4.2.2). Thus the thiuram disulphides (LXXIV) and the benzthiazyl sulphenamides (LXXV) both give the peroxidolytic antioxidants XXXIV and XXXV whose antioxidant

$$\underset{\text{LXXIV}}{R_2NC\overset{\overset{S}{\|}}{}-S-S-\overset{\overset{S}{\|}}{C}NR_2} \xrightarrow{\text{ZnO}} \underset{\text{(XXXIV)}}{R_2NC\overset{S}{\underset{S}{\diagdown}}Zn\overset{S}{\underset{S}{\diagdown}}CNR_2} \tag{5.5}$$

$$2\ \underset{\text{LXXV}}{\left[\text{benzothiazyl}\right]C-S-NHR} \xrightarrow{\text{ZnO/S}} \underset{\text{(XXXV)}}{\left[\left[\text{benzothiazyl}\right]C-S\right]_2 Zn} \tag{5.6}$$

mechanisms were discussed in section 5.2.3. Most disulphide and sulphenamide accelerators behave similarly. Rubbers produced in this way have a number of disadvantages. The most important of these is that they are generally costly to produce and their technological properties are frequently inferior to rubbers produced with more conventional vulcanisation systems.

An effective rubber antidegradant has to perform a variety of functions. Not only has it to protect the rubber against the effects of oxidation at high temperatures but it must also prevent mechano-oxidative breakdown (see chapter 4, section 4.2.5). Only one class of antioxidants, the 4-alkylaminodiphenylamines, of which isopropyl-phenyl-p-phenylene diamine, IPPD, I(b), is the best known commercial example, has proved to be effective in all conditions.

$$\underset{\text{I}(b)}{\text{(phenyl)}-NH-\text{(phenylene)}-NH^{i}Pr}$$

As a heat stabiliser IPPD functions as a CB-D antioxidant (see scheme 5.4). However, in rubbers under conditions of stress, the rate of formation of macro-alkyl radicals in much higher than in thermal oxidation, and under these conditions, the CB-A/CB-D catalytic cycle (scheme 5.3) involving the derived nitroxyl radical IV assumes importance. After an initial rise and fall the concentration of the nitroxyl remains stationary during most of the lifetime of the polymer (see Fig. 5.8) and the presence of the corresponding hydroxylamine (LXXVI) indicates the operation of the cyclical process shown in scheme 5.17. As might be expected, the nitroxyl radical is somewhat more effective than the parent amine (see table 5.8). Other secondary arylamines and their oxidation products behave similarly, and galvinoxyl (XXIII), although not as effective as the diaryl nitroxyl radicals, is also an antifatigue agent.

The antiozonant mechanism of the p-phenylene diamines is less well understood than their antifatigue activity. They are known to react very rapidly with ozone, and it may be that preferential reaction with ozone in the surface of rubbers is one of the functions of an antiozonant. However, there is evidence that the reaction products of ozone and p-phenylene diamines form a coherent ozone-unreactive film on the surface of the rubber and a physical protective role seems to be another possibility.

5.3.3. Polymer-bound antioxidants

Rubbers and plastics are being increasingly used in environments which are hostile to their long-term durability. High temperatures lead to the rapid

Figure 5.8. Kinetics of formation and decay of alkylperoxyl and nitroxyl radicals in rubber during flexing. *A* alkylperoxyl; *B* nitroxyl. (Reproduced by kind permission of *Chem. & Ind.*, p. 573, 1980.)

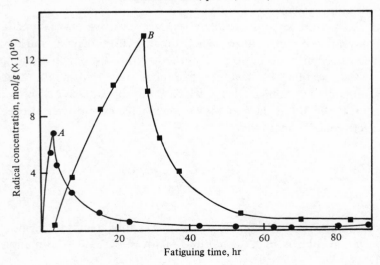

Scheme 5.17 Catalytic antioxidant action of IPPD (1*b*) during its function as an anti-fatigue agent in rubber.

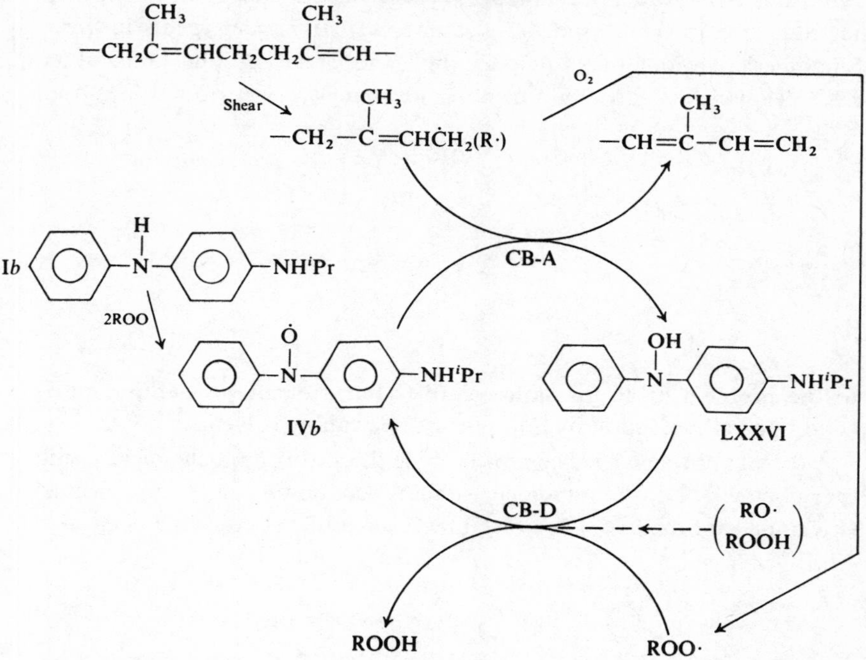

Table 5.8. *Antifatigue activity in rubbers of secondary arylamines (I) and their derived oxidation products (concentration 1 g/100 g).*

	Hours to failure[+]		
Antioxidant	X = H	X = O·	X = OH
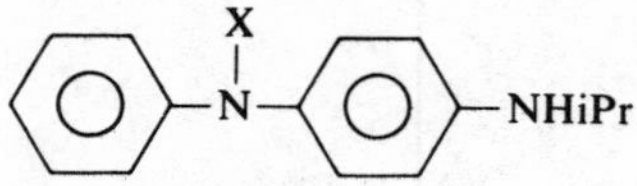 ⬡–N(X)–⬡–NHiPr	274	337	—
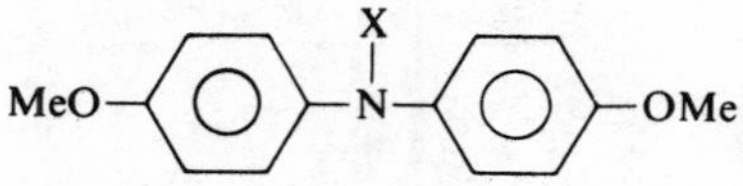 MeO–⬡–N(X)–⬡–OMe	207	441	56*
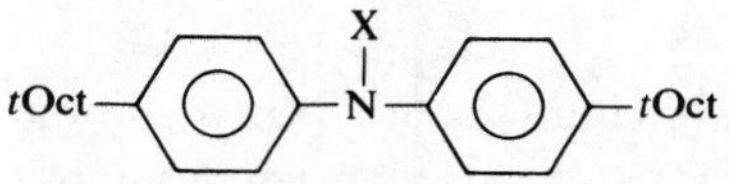 tOct–⬡–N(X)–⬡–tOct	83	88	107

Control (no additive), 20 hours
[+] In a Monsanto fatigue-to-failure machine.
* Additive insoluble, lost rapidly from the surface.

loss of many antioxidants and stabilisers from polymers, but equally important is the leaching action of aggressive media such as lubricating oils in engine components, and dry cleaning solvents or detergents in fibres. Oligomeric antioxidants such as the commercial product Flectol H, LXXVII, which is partially polymeric, go some way toward a solution

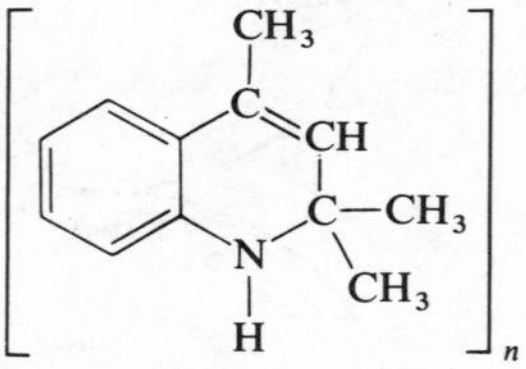

LXXVII

to the problem of antioxidant volatility but are not substantive under conditions of extraction by hot oils or dry cleaning solvents.

A recent approach has been to combine the antioxidant chemically with the polymer. Nitrile-butadiene rubbers containing a copolymerised vinylamide antioxidant, LXXVIII, have been found to resist the aggressive

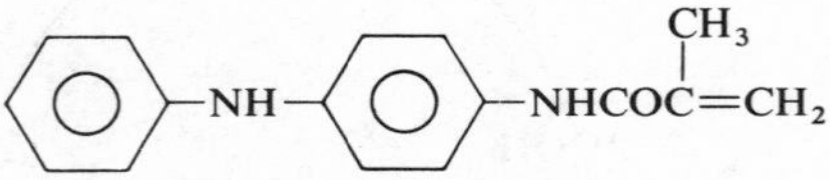

LXXVIII

conditions experienced by an engine oil seal much more successfully than additives such as LXXVII. Similarly, reaction of rubbers after manufacture in the latex or during processing with thiol antioxidants, LXXIX or LXXX, leads to a similar resistance to oil extraction.

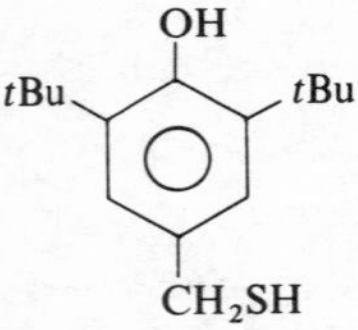

LXXIX

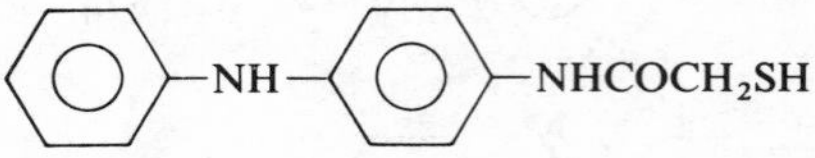

LXXX

The reaction involved is the classical peroxide catalysed addition to double bonds (see scheme 5.18), and it is applicable to any polymer containing double bonds including rubber modified plastics (e.g. acrylonitrile-butadiene-styrene copolymers, ABS). Figure 5.9 compares the behaviour of a nitrile rubber adduct of LXXX (MADA-B),

Scheme 5.18 Formation of thiol antioxidant adducts in natural rubber.

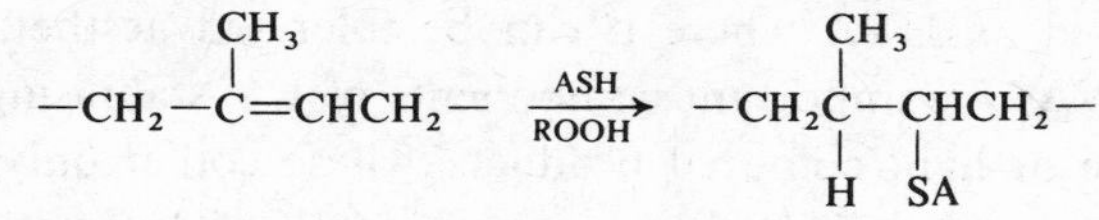

$$-CH_2-\underset{\underset{CH_3}{|}}{\overset{\overset{CH_3}{|}}{C}}=CHCH_2- \xrightarrow[ROOH]{ASH} -CH_2\underset{\underset{H}{|}}{\overset{\overset{CH_3}{|}}{C}}-\underset{\underset{SA}{|}}{CH}CH_2-$$

ASH = LXXIX or LXXX

with the oligomeric antioxidant (LXXVII) in a cyclical hot oil/hot air test at 150 °C. Nitrile rubber normally undergoes rapid hardening under these conditions so that even after one cycle, it is unable to perform its technological function in an engine seal. Figure 5.9 shows that the oxidation process which leads to the phenomenon is effectively retarded in the rubber protected by bound antioxidant whereas the conventional anti-oxidant, Flectol H (LXXVII), is relatively ineffective.

A potentially important use for polymer bound antioxidants is in car and truck tyres. Conventional antidegradants such as IPPD (I*b*) (see section 5.3.2) are leached from tyres by road surface water during the life of a tyre. Consequently, retreaded tyres may be inadequately protected against high-temperature fatigue and, as the trend toward retreading increases, the use of more substantive antioxidants is seen to be increasingly necessary.

5.3.4. UV stabilisers

The earliest UV stabilisers to be developed were UV absorbers (UVAs) with a high molar extinction coefficient at the shortest wavelength of the Sun's

Figure 5.9. Effect of a cyclical hot oil (150 °C for 24 h)/hot air (150 °C for 24 h) test on the hardness of nitrile–butadiene rubber containing macromolecular antioxidants. 1 control, without antioxidant; 2 Flectol H, (LXXVII); 3 Chemigum HR665, (copolymerised LXXVIII); 4 MADA-B (adduct of LXXX). All antioxidants at 2 g/100 g concentration.

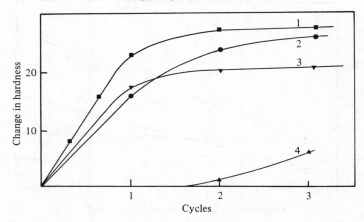

spectral region (290–350 nm). Carbon black absorbs all wavelengths and is still widely used as UVA where it can be tolerated aesthetically. The 'colourless' UVAs are important components of UV-stabilising formulations for white or light coloured products. These confer only moderate protection when used alone but in synergistic combinations with both CB-D and PD-C antioxidants they can be highly effective stabilisers for hydrocarbon polymers.

Polyolefins

Most of the published work on the photo-stabilisation of polymers is concerned with this group, particularly when used in fibres. The other fibre-forming polymers, the polyesters, polyamides, polyacrylonitrile and cellulosics are relatively photo-stable compared with the polyolefins (see section 4.1.5) and this fact justifies the emphasis currently devoted to the photo-stabilisation of the hydrocarbon polymers.

Figure 5.10 compares the effectiveness of some typical commercial antioxidants and UV stabilisers acting by a variety of mechanisms as a function of their concentration in polypropylene. Irganox 1076 (VIII, appendix 5(i)) and ZnDEC (zinc diethyldithiocarbamate, XXXIV, R = Et, M = Zn) are typical thermal antioxidants, the former acting by the CB-D mechanism and the latter by the PD-C mechanism. Both are UV stabilisers at low concentration but their effectiveness increases little with concentration. At

Figure 5.10. Effect of additives on the UV stability of polypropylene. ▼ Tinuvin 770; ■ NiDEC; □ HOBP; ▲ Irganox 1076; ● ZnDEC. (Reproduced by kind permission of *Pure and Appl. Chem.*, **52**, p. 365, 1980.)

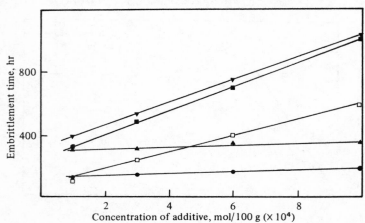

the concentrations used in practice, therefore, they are much less effective than 2-hydroxy-4-octoxybenzophenone (HOBP, Cyasorb UV 531, XLIII), a typical UV absorber, in spite of the fact that they are more effective at lower concentrations. HOBP does not function only as a UV screen, however. It appears in addition to behave as a weak CB-D antioxidant, but unlike the hindered phenols, it is quite resistant to photolysis. In heavily oxidised polymers, one of the functions of HOBP is to remove excited carbonyl species by hydrogen transfer (CB-D) and it seems likely that the benzotriazoles (XLIX) behave similarly.

Nickel diethyldithiocarbamate (NiDEC, XXXIV, R = Et, M = Ni) is, like ZnDEC, an effective PD-C antioxidant but it resembles HOBP in that it absorbs light strongly in the 330 nm region of the spectrum and is consequently very photo-stable. Figure 5.11 shows that the induction period to photo-oxidation of the polymer is directly related to the time taken for the metal complexes to be destroyed.

The combined use of an antioxidant (either CB or PD) and a UVA leads to synergism (see table 5.9). The role of the UVA is primarily to protect the antioxidant from photo-oxidation under service conditions. However, a complementary reason for the synergistic effect is that the antioxidants also protect the UV absorber from oxidation by hydroperoxides during processing. It will be evident from the foregoing discussion that effective UV stabilisers for polymers are essentially UV-stable antioxidants.

It was seen in earlier sections that the metal dithiocarbamates are

Figure 5.11. Correlation of the decay of UV absorbance of ZnDEC (285 nm) and NiDEC (330 nm) in low-density polyethylene with photo-oxidation induction period. Initial additive concentration, 3×10^{-4} mol/100 g. (Reproduced by kind permission of *Pure & Appl. Chem.*, **52**, p. 365, 1980.)

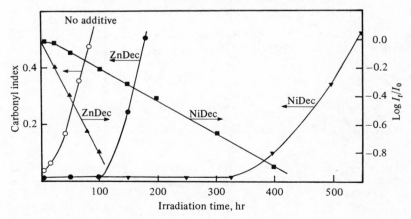

effective melt stabilisers during the processing of polymers and thermal antioxidants in use. Antioxidant activity is, however, associated with the ligand rather than the metal ion and indeed it is remarkable that compounds containing metal ions can exert such a powerful antioxidant effect in view of the known tendency for transition metal ions to catalyse both thermal and photo-oxidation reactions in polymers. It has also been seen (figure 5.11) that once a photo-stabilising metal complex has been destroyed by the photo-oxidative reactions occurring in the polymer, it ceases to have any antioxidant activity and the polymer is destroyed by oxidation. This principle has been extended in the design of photo-degradable plastics with a built-in time control mechanism.

The iron dialkyldithiocarbamates (XXIV, $M = Fe(III)$), are very much less stable to light than the Ni(II) or Co(III) complexes. In low concentrations iron dimethyldithiocarbamate (FeDMC, XXIV, $M = Fe(III)$, $R = Me$) is a melt stabiliser but photo-activator for the photo-oxidation of polyolefins. At higher concentrations, however, it gives rise to a photo-induction period similar to but shorter than the nickel complex (Fig. 5.11) and at the end of the induction period the liberated ionic iron results in a much more rapid rate of photo-oxidation than occurs in its absence. Thus, by varying the concentration it is possible to vary the lifetime of the polymer controllably over a wide rime scale. A development of this principle is used commercially. This involves a combination of a stabilising nickel dithio-carbamate and an activating iron dithiocarbamate which allows the lifetime of polyolefins to be varied controllably on both sides of the natural lifetime of the polymer by a factor of almost 25. The presence of the iron dithiocarbamate ensures that the degradation is rapid and complete at the

Table 5.9. *Synergism between antioxidants and a UV absorber in polypropylene (all additives at 0.2 g/100 g).*

Antioxidant	UV Absorber	Embrittlement time (hours)
—	—	70
ZnDNC	—	385
Irganox 1076	—	325
NiDNC	—	1160
—	UV 531	400
ZnDNC	UV 531	1400
NiDNC	UV 531	2250
Irganox 1076	UV 531	1250

end of the induction period; an important requirement for plastic waste and litter (see chapter 1, section 1.5).

Several commercial stabilisers listed in appendix 5(ii) have a similar dual mechanistic role to the dithiocarbamates. Three examples of UV-stable CB-D antioxidants are Cyasorb 1084 (L), Tinuvin 120 (LXXXI) and Irgastab 2002 (LXXXII). L and LXXXII are UV stable by virtue of the strongly absorbing nickel complex and LXXXI because of the presence of the benzoate ester.

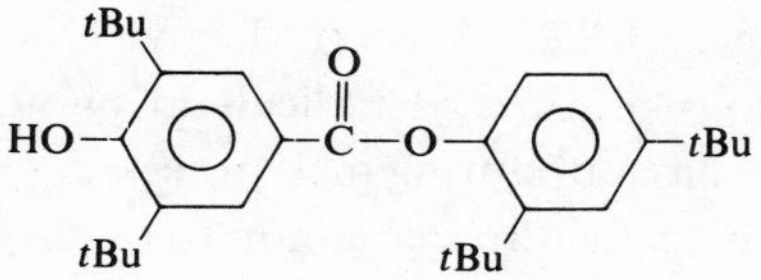

LXXXI

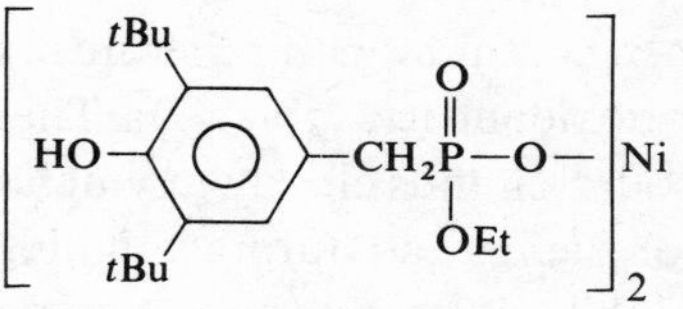

LXXXII

The hindered piperidines, of which Tinuvin 770 (LXXXIIIa) is the most important commercial example, are unique among the effective UV stabilisers in that they do not absorb UV light. Moreover, they are rapidly transformed, partially during processing and completely during the early

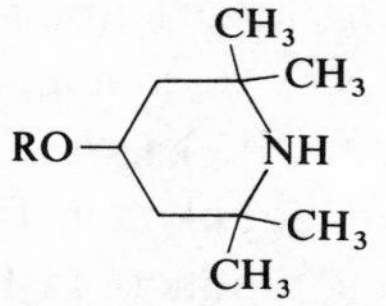

LXXXIII

LXXXIIIa, R = HN

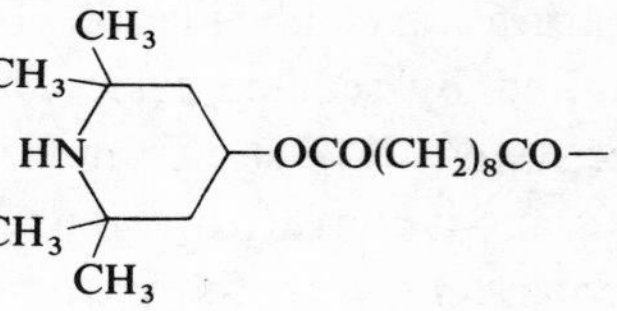

Table 5.10. *Effectiveness of a hindered piperidine and its derived oxidation products as UV stabilisers for polypropylene.*

	Embrittlement time (hours)
Unstabilised control	90
LXXXIII	750
LXXXIV	920
LXXXV	1040

stages of photo-oxidation, to the corresponding nitroxyl radical (LXXXIV) which is even more effective than the parent amine as a UV stabiliser (see table 5.10).

$$
\begin{array}{cc}
\text{LXXXIV} & \text{LXXXV} \\
\end{array}
$$

Structure LXXXIV: RO group attached to a piperidine ring with N—O· and CH$_3$ groups. Structure LXXXV: RO group attached to a piperidine ring with N—OH and CH$_3$ groups.

In common with other nitroxyl radicals, LXXXIV is an effective CB-A antioxidant (cf. table 5.3). The behaviour of nitroxyl radicals in photo-oxidation thus closely resembles their behaviour during melt processing of polyolefins and during fatiguing of rubbers. Features common to all these inhibition processes are a high rate of macro-radical formation and restricted access of oxygen. In the case of a solid polymer undergoing initiated oxidation, the ratio $[R\cdot]/[ROO\cdot]$ may be several orders of magnitude higher than it is in a liquid hydrocarbon of similar chemical structure, and this fact provides the necessary conditions for the operation of the CB-A/CB-D catalytic cycle outlined generally in scheme 5.3 and elaborated for photo-oxidation in scheme 5.19. In the present case, the alkyl hydroxylamine (LXXXVI), free hydroxylamine (LXXXV) and olefinic unsaturation can be identified during the course of the photo-oxidation as required by scheme 5.19. The hydroxylamine (LXXXV) is even more effective as a UV stabiliser than the nitroxyl radical (LXXXIV), (see table 5.10), and is converted to it under oxidative conditions in the polymer. This is consistent with the participation of these oxidation products in the cyclical mechanism. The role of the alkyl hydroxylamine is less clear. It is known to regenerate nitroxyl radicals rapidly in the presence of oxygen, but

Table 5.11. *Antagonism between a hindered piperidine and a nickel dithiocarbamate in the UV stabilisation of LDPE* (*additives all at 3×10^{-4} mole/100 g*).

Antioxidant	Embrittlement time (hours)
None	1200
NiDEC	1800
Tinuvin 770	2250
NiDEC + Tinuvin 770	1550

Scheme 5.19 Catalytic role of hindered piperidinoxyl radicals during the photostabilisation of polypropylene.

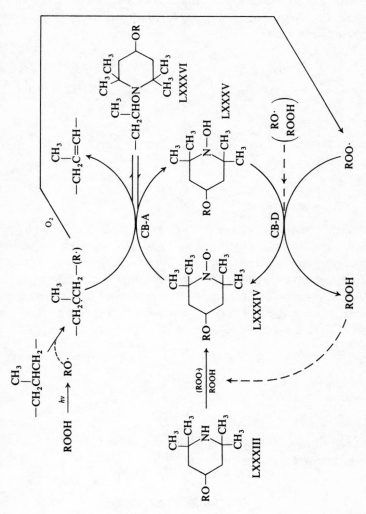

it seems probable that this simply reflects the ease of dissociation of the alkyl hydroxylamine to the parent radicals in a non-terminating process (reaction 5.7). Termination must involve the removal of free radicals in the catalytic cycle (scheme 5.19). Hydroperoxides and their radical dissociation products can also readily react with the free hydroxylamine and cannot be detected during the induction period. This contributes to the regeneration of the nitroxyl radical.

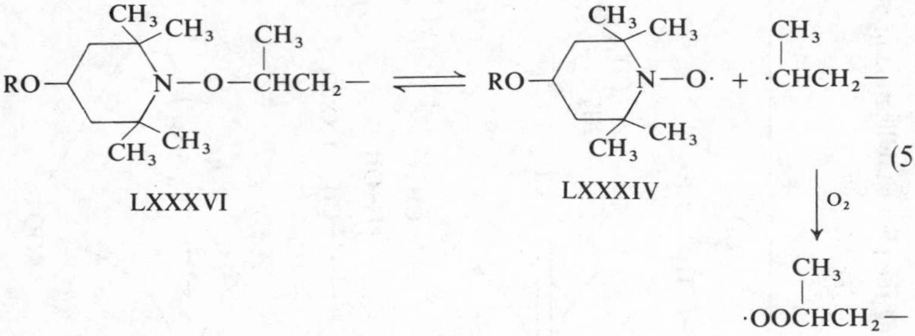

$$\text{LXXXVI} \rightleftharpoons \text{LXXXIV} \qquad (5.7)$$

Although hindered piperidines show co-operative effects with UVAs, they are antagonistic toward the peroxidolytic UV stabilisers. This is shown typically in polyethylene in table 5.11. There is evidence that the nitroxyl radicals are primarily formed from the hindered piperidines by reaction with hydroperoxides (reaction 5.8), and it has been suggested that by interfering with the first stage of this process peroxidolytic antioxidants reduce the concentration of nitroxyl in the polymer.

$$>\!N\!-\!H + ROOH \longrightarrow >\!N\!\cdot + RO\!\cdot + H_2O$$

$$\Big\downarrow \begin{array}{c} ROO\cdot \\ or\ O_2 \end{array}$$

$$>\!N\!-\!O\!\cdot + RO\!\cdot \longleftarrow [>\!N\!-\!OOR] \qquad (5.8)$$

However, irreversible trapping of nitroxyl radicals by thiyl radicals may also be involved.

Suggested further reading

1. G. Scott, Mechanisms of Antioxidant Action, *Developments in Polymer Stabilisation-4*, ed. G. Scott, App. Sci. Pub., London, 1981, chapter 1.
2. J. Pospisil, Chain-Breaking Antioxidants in Polymer Stabilisation, *Developments in Polymer Stabilisation-1*, ed. G. Scott, App. Sci. Pub., London, 1979, chapter 1.
3. T. J. Henman, Melt Stabilisation of Polypropylene, *Developments in Polymer Stabilisation-1*, ed. G. Scott, App. Sci. Pub., London, 1979, chapter 2.

4. B. B. Cooray and G. Scott, The Role of Tin Stabilisers in the Processing and Service Performance of PVC, *Developments in Polymer Stabilisation-2*, ed. G. Scott, App. Sci. Pub., London, 1980, chapter 2.

5. S. Al-Malaika and G. Scott, Thermal Stabilisation of Polyolefins, *Degradation and Stabilisation of Polyolefins*, ed. N. S. Allen, App. Sci. Pub., London, 1983, chapter 6.

6. S. Al-Malaika and G. Scott, Photo-stabilisation of Polyolefins, *Degradation and Stabilisation of Polyolefins*, ed. N. S. Allen, App. Sci. Pub., London, 1983, chapter 7.

7. G. Scott, Peroxidolytic Antioxidants, Sulphur Antioxidants and Autosynergistic Stabilisers Based on Alkyl and Aryl Sulphides, *Developments in Polymer Stabilisation-6*, ed. G. Scott, App. Sci. Pub., London, 1983, chapter 2.

8. S. Al-Malaika, K. B. Chakraborty and G. Scott, Peroxidolytic Antioxidants, Metal Complexes Containing Sulphur Ligands, *Developments in Polymer Stabilisation-6*, ed. G. Scott, App. Sci. Pub., London, 1983, chapter 3.

9. G. Scott, Stable Radicals as Antioxidants in Polymers, *Developments in Polymer Stabilisation-7*, ed. G. Scott, App. Sci. Pub., London, 1984, Chapter 2.

10. J. Pospisil, Aromatic Amine Antidegradants, *Developments in Polymer Stabilisation-7*, ed. G. Scott, App. Pub. London, 1984, Chapter 1.

11. A. Guyot and A. Michel, Stabilisation of Polyvinyl Chloride with Metal Soaps and Organic Compounds, *Developments in Polymer Stabilisation-2*, ed. G. Scott, App. Sci. Pub., London, 1980, chapter 3.

12. N. C. Billingham and P. D. Calvert, The Physical Chemistry of Oxidation and Stabilisation of Polyolefins, *Developments in Polymer Stabilisation-3*, ed. G. Scott, App. Sci. Pub., London, 1980, chapter 5.

13. D. Gilead and G. Scott, Time-Controlled Stabilisation of Polymers, *Developments in Polymer Stabilisation-5*, ed. G. Scott, App. Sci. Pub., London, 1982, chapter 4.

14. V. Yu. Shlyapintokh and V. B. Ivanov, Antioxidant Action of Sterically Hindered Amines and Related Compounds, *Developments in Polymer Stabilisation-5*, ed. G. Scott, App. Sci. Pub., London, 1982, chapter 3.

15. F. Tudos, G. Balint and T. Kelen, The Effect of Various photo-stabilisers on the Photo-oxidation of Polypropylene, *Developments in Polymer Stabilisation-6*, ed. G. Scott, App. Sci. Pub., London, 1983, chapter 4.

16. D. J. Carlsson, A. Garton and D. M. Wiles, The Photo-Stabilisation of Polyolefins, *Developments in Polymer Stabilisation-1*, ed. G. Scott, App. Sci. Pub., London 1979, chapter 7.

Chemical structures of commercial antioxidants referred to in the text

Registered trade name	Code used in text	Structure
(A) Arylamines (CB-D)		
Nonox OD Naugawhite	I(a)	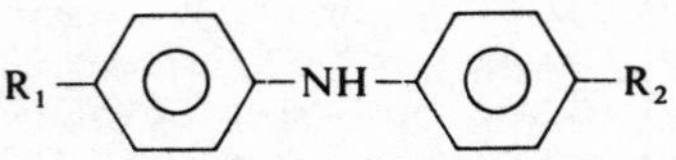 $R_1 = R_2 = {}^tOct$
Santoflex IP Nonox ZA	I(b) IPPD	$R_1 = H, R_2 = NH^iPr$
Nonox DPPD	I(c) DPPD	$R_1 = H, R_2 = NHPh$
Flectol H	LXXVII	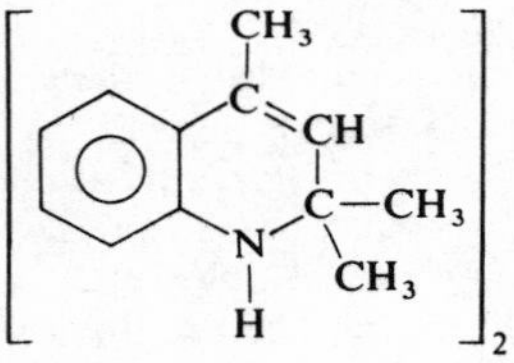
(B) Phenols (CB-D)		
Topanol OC Ionol	VI BHT	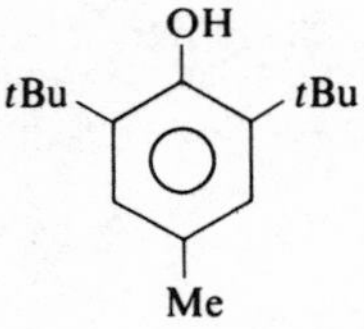
Anti-oxidant 2246	VII	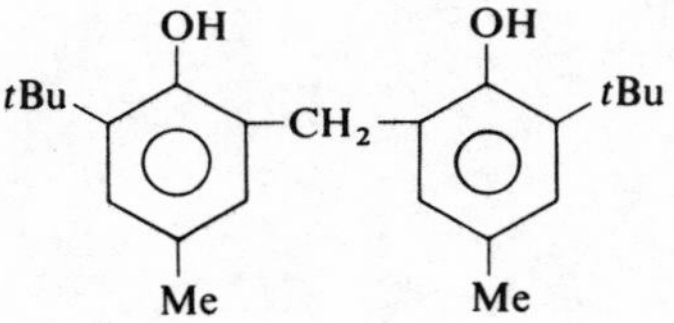
Irganox 1076	VIII	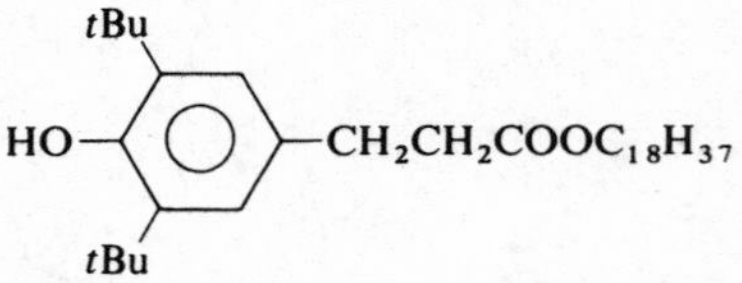

Registered trade name	Code used in text	Structure
Irganox 1010	IX	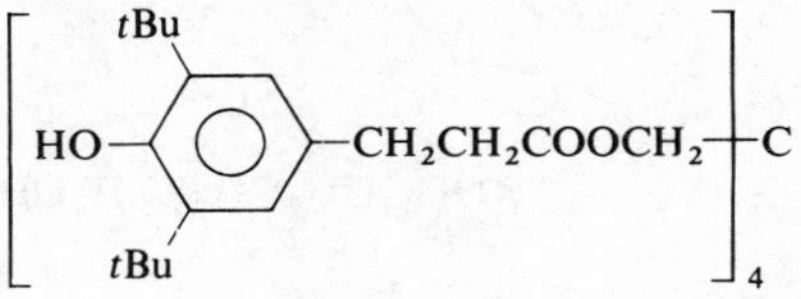
Ethyl 330	X	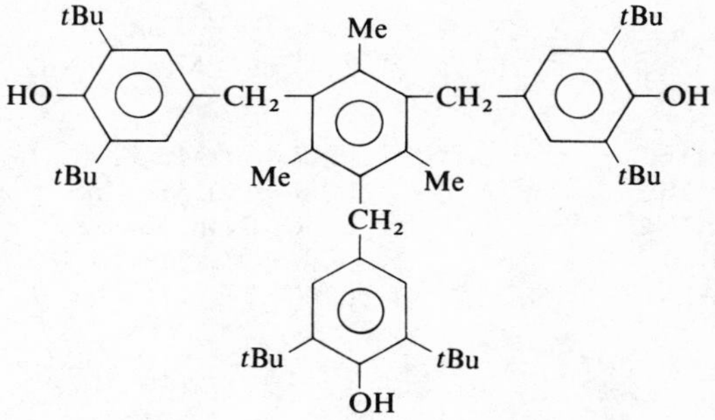
Goodrite 3114	XI	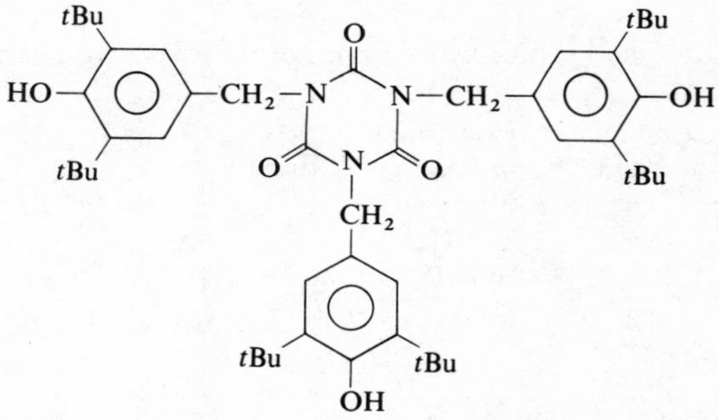
Topanol CA	XII	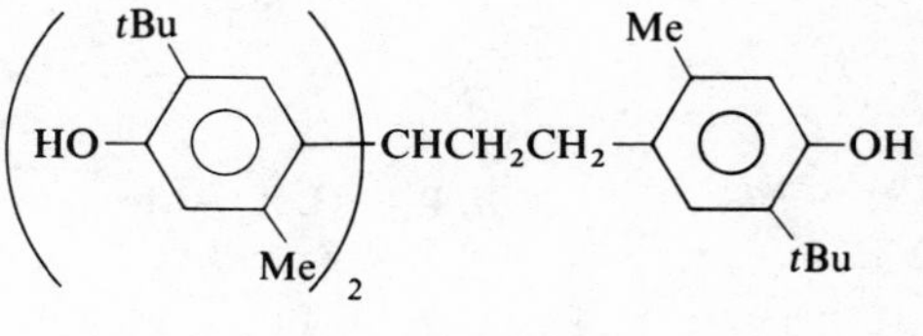
(C) Phosphites (PD-S)		
Naugard	XXV	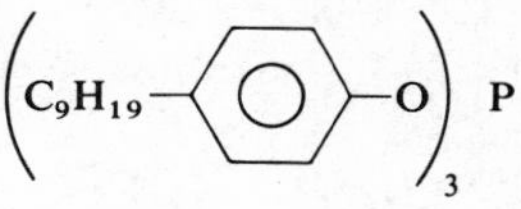

Registered trade name	Code used in text	Structure
(D) Sulphur compounds (PD-C)		
	XXVI	$ROCOCH_2CH_2SCH_2CH_2COOR$
DLTP		(a) $R = C_{12}H_{25}$
DSTP		(b) $R = C_{18}H_{37}$
	XXXIV	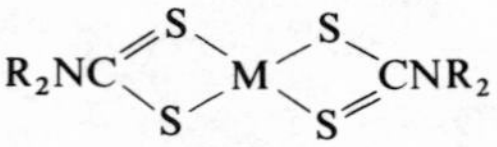
ZnDEC		(a) R = Et, M = Zn
ZnDBC		(b) R = ^{n}Bu, M = Zn
ZnDNC		(c) R = ^{i}Non, M = Zn
NiDBC		(d) R = ^{n}Bu, M = Ni
NiDNC		(e) R = ^{i}Non, M = Ni

Appendix 5(ii)

Chemical structure of commercial UV stabilisers referred to with text

Registered trade name	Code used in text	Structure
Cyasorb UV 531	XLVIII HOBP	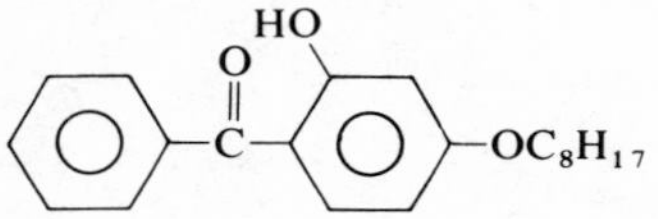
	XLIX	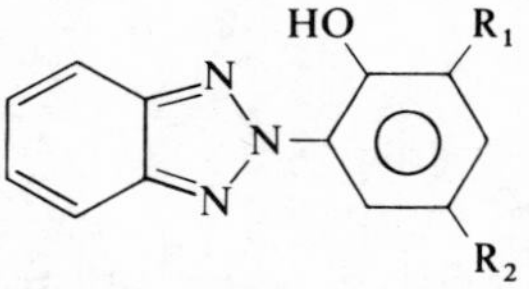
Tinuvin P		(a) $R_1 = H, R_2 = Me$
Tinuvin 327		(b) $R_1 = R_2 = {}^tBu$
Cyasorb 1084	L	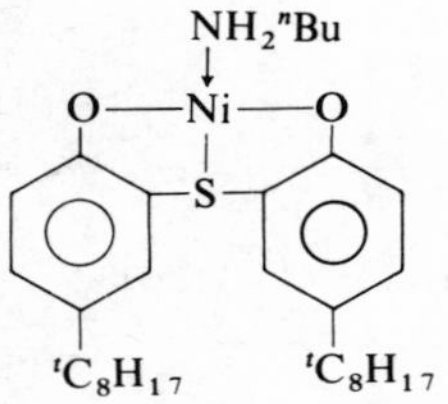

Registered trade name	Code used in text	Structure
Tinuvin 120	LXXXI	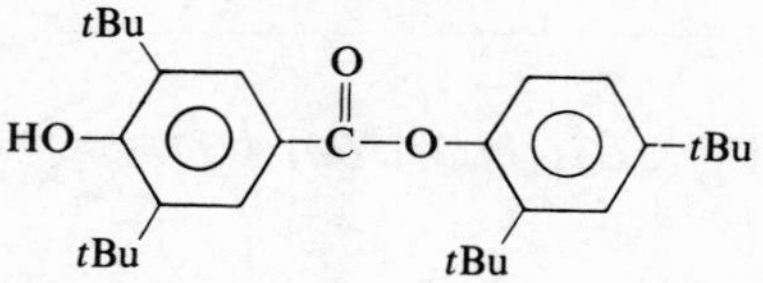
Irgastab 2002	LXXXII	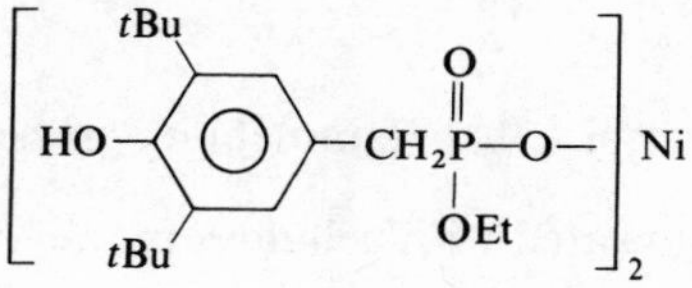
Tinuvin 770	LXXXIII (*a*)	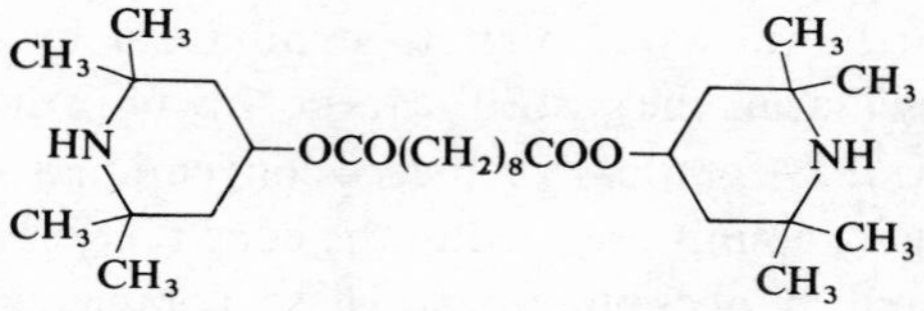

6

Degradation and the fire hazard

6.1. The flammability problem

The fire hazard associated with cellulose in the form of wood, textiles and paper has caused concern from the earliest times and attempts have been made for hundreds of years to minimize it by the use of fire retardant additives. During the past 30 years polymers, in the form of rubbers, plastics and fibres, have gradually replaced conventional materials until, at present, very large amounts are used in the construction and furnishing of homes, offices and public buildings. These commercial polymers are almost exclusively organic compounds which are frequently even more flammable than the conventional cellulosic materials and are thus associated with a much increased fire hazard.

The flammability of polymers and the associated destruction of property is not the only problem, however. Fatalities in fire incidents are not normally the result of burning. Instead, victims are suffocated by smoke or poisoned by noxious gases. A secondary effect of smoke is to limit visibility thereby making escape more difficult which in turn leads to panic. Unfortunately, compared with conventional materials, the smoke from plastics is often more dense and the fumes more poisonous. In addition, the normal methods of reducing flammability usually increase the density of the smoke and fumes.

The almost inevitable presence of synthetic polymeric materials in burning buildings also introduces new problems in fire fighting. Before the advent of these synthetics, a well-trained fire fighter could make a very quick and accurate assessment of the smoke and fume hazard on entering a burning building knowing that the principal poisonous material was liable to be carbon monoxide. He knew almost instinctively when it became necessary to wear breathing apparatus to counteract these hazards. However, the burning characteristics of synthetic polymeric materials are so varied and the smoke may contain such a variety of dangerously toxic materials that he must almost automatically wear breathing apparatus with a consequent loss of fire fighting efficiency.

Thus the problems involved in reducing the fire hazards associated with synthetic materials are very complex. They are not only scientific but also economic and social. The relative cheapness of modern materials and their contribution to the improvement of the quality of life must be balanced against the increased twin hazards of loss of property by combustion and flame and the loss of human life by smoke and fumes. Indeed decisions about the application of alternative materials will often involve a decision about the relative importance of these twin hazards.

6.2. Flammability testing

The fact that the description 'fire retardant' is now invariably used in preference to 'fire proof' is a clear indication that few people now believe that synthetic polymers may ever be rendered absolutely non-flammable by appropriate treatment. In order to monitor progress towards achieving optimum fire retardant properties in a commercial material it is clearly desirable to have a quantitative measure of flammability. This leads to the question of what is meant by flammability and it is not difficult to appreciate that any definition will depend very much upon one's immediate point of view. For example, there are methods of estimating flammability which depend upon ease of ignition, rate of flame spread and duration of burning after ignition (self-extinction time). On the other hand, the rate of generation and density of smoke may be important or the toxicity of the fumes.

The situation is further complicated by the fact that all of these properties are affected not only by the chemical nature of the material but also by secondary factors such as the temperature, rate of flow and composition of the surrounding atmosphere, as well as the geometry of the article being tested and the way in which it is burned. For example, a strip of material may be placed vertically, horizontally, or in any intermediate configuration and may be ignited by the application of a flame to the top or bottom or even by radiant heat. Clearly there can be no unique quantitative measure of flammability.

One very well established and standardised measure of flammability is the Limited Oxygen Index (LOI) or Candle Test, in which a strip of the material under test is placed vertically in a flowing atmosphere of oxygen and nitrogen and ignited at the top. The composition of the atmosphere is measured at which burning is just maintained and the LOI is given by the mole fraction of oxygen in the mixture, usually expressed as a percentage.

$$LOI = \frac{[O_2]}{[O_2]+[N_2]} \times 100 \qquad (i)$$

Some typical values are presented in table 6.1. Since air contains a mole fraction of oxygen of approximately 0.21, it is clear that some of the commonest plastics will continue to burn in air after ignition. It is of interest to notice that those which contain chlorine are very much less flammable and polytetrafluoroethylene requires almost pure oxygen. This test gives only very limited information. For example, it gives none at all about rate of spread of flame or burning properties and indeed it is well known that under 'real fire' conditions materials with LOI values much greater than 21% can still be dangerously flammable. It is also clear that tests must be specific to the article being burned. Thus, for example, nylon carpet will give different results from nylon curtains and clothing.

Thus while simple tests of this kind can give limited information about flammability it seems that the only ultimately satisfactory test method must involve the full scale reproduction of a real fire situation with, for example, real furniture and furnishings disposed as in a room or office and with an accurately simulated system of airflow and ventilation. It has been demonstrated that even the architecture and detailed mode of construction of the articles of furniture may play a vital role in their flammability.

6.3. The burning cycle

Strictly speaking, polymers do not burn. It is the flammable volatile products of decomposition which do so. Thus the combustion of a polymer,

Table 6.1. *Values of limiting oxygen index (LOI) for some commercial polymers.*

Material	LOI
polyoxymethylene	16
poly(methyl methacrylate)	17
polyethylene	17
polypropylene	17
polystyrene	18
poly(vinyl alcohol)	22
nylon	25
poly(phenylene oxide)	30
poly(vinyl chloride)	47
poly(vinylidene chloride)	60
polytetrafluoroethylene	95

or indeed of any solid material, may be represented as in scheme 6.1.

Scheme 6.1 A model of the burning process.

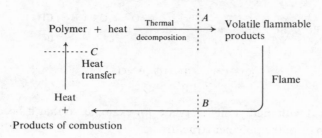

For continuous burning to occur, therefore, the application of heat must be sufficient to decompose the material, the temperature must be high enough to ignite these products of degradation, and the amount of heat transferred back to the polymer must be great enough to maintain the cycle when the initially applied source of heat is withdrawn. It is clear, therefore, in principle at least, that this cycle must be broken at one or more of the points *A*, *B* and *C* if a polymer is to be made fire retarded. The solution of a fire retardance problem will usually involve one of the following: (*a*) modification of the thermal degradation process; (*b*) quenching the flame; (*c*) reduction of the supply of heat from the flame back to the decomposing polymer. It will become clear in the following pages that all of these are closely associated with polymer degradation processes and this is the justification for including this discussion of fire retardance in the present volume. It should be stated at the outset, however, that most fire retardants have been discovered by empirical methods and their mode of action is understood only in the most general terms. Nevertheless there is no doubt that most fire retardants act in more than one way.

6.4. 'Additives' and 'reactives'

Fire-retardant chemicals may be incorporated into polymeric materials in two ways – as 'additives' and as 'reactives'. Additives are compounds which are mechanically mixed with the material to be protected, usually during the processing sequence. Reactives, on the other hand, are materials which are chemically bound as an integral part of the polymeric structure.

Bromine, for example, is an effective fire retarding element which may be incorporated in an additive like decabromobiphenyl (I) or in a reactive like 2-dibromoethyl methacrylate (II) which may be

$$\text{I}$$

Structure I: brominated biphenyl with Br substituents.

$$\begin{array}{c} CH_3 \\ | \\ CH_2{=}C \\ | \\ C \\ O{\diagup}\quad{\diagdown}O{-}CH_2{-}CH_2 \\ \qquad\qquad\quad | \\ \qquad\qquad\quad Br \end{array}$$

$$\text{II}$$

$$(Cl{-}CH_2{-}CH_2{-}O)_3 P{=}O$$
$$\text{III}$$

copolymerised with methyl methacrylate, for example, so that it becomes an integral part of the polymer structure

$$\begin{array}{ccccccc} & CH_3 & & CH_3 & & & CH_3 \\ & | & & | & & & | \\ {\sim}CH_2{-}C{-}CH_2{-}{-}C{-}CH_2{-}{-}{-}{-} & & {-}C{\sim} \\ & | & & | & & & | \\ & C & & C & & & C \\ O{\diagup}{\diagdown}OCH_3 & O{\diagup}{\diagdown}OCH_2{-}CH_2 & O{\diagup}{\diagdown}OCH_3 \\ & & & \qquad\quad| \\ & & & \qquad\quad Br \end{array}$$

Similarly, phosphorus, which is also an important fire retarding element, may be incorporated into polyurethane in the form of an additive like *tris*-(2-chloroethyl)phosphate (III) or as a reactive in the form of a phosphorus containing diol which may be reacted with diisocyanate in the polymerisation process, reaction (6.1).

$$HO{-}\overset{\overset{\textstyle O}{\|}}{\underset{\underset{\textstyle \phi}{|}}{P}}{-}O{-}(CH_2)_2{-}O{-}\overset{\overset{\textstyle O}{\|}}{\underset{\underset{\textstyle \phi}{|}}{P}}{-}OH \; + \; OCN{-}\langle\bigcirc\rangle{-}NCO \longrightarrow$$

$${\sim}O{-}\overset{\overset{\textstyle O}{\|}}{\underset{\underset{\textstyle \phi}{|}}{P}}{-}O{-}(CH_2)_2{-}O{-}\overset{\overset{\textstyle O}{\|}}{\underset{\underset{\textstyle \phi}{|}}{P}}{-}O{-}\overset{\overset{\textstyle O}{\|}}{C}{-}\overset{\overset{\textstyle H}{|}}{N}{-}\langle\bigcirc\rangle{-}\overset{\overset{\textstyle H}{|}}{N}{-}\overset{\overset{\textstyle O}{\|}}{C}{\sim}$$

$$(6.1)$$

Reactives have the advantage over additives that, as in the case of bound antioxidants discussed in the previous chapter, they cannot be lost by evaporation or leaching with water or other solvents during the useful life of the polymer. In addition they should be expected to be immediately available to protect the polymer in a fire situation since their liberation in the polymer will be simultaneous with the decomposition of the polymer while the decomposition or volatilisation temperature of an additive must be carefully matched to that of the polymer so that it will not become available before it is required or remain unchanged while the polymer is decomposing. This 'right place at the right time' principle is vitally important in fire protection.

On the other hand, one possible disadvantage of reactives over additives is that, being incorporated into the polymer structure, they are much more likely to affect the chemical stability of the polymer. It is also very important that both additives and reactives must be stable at processing temperatures.

The desirable physical properties of polymers which make them valuable in specific applications are of course highly dependent upon their chemical structures. Consequently, flame retardants must not be allowed to significantly affect mechanical strength, electrical and optical properties or susceptibility to oxidation and photo- and thermal breakdown. The choice of fire retardant for a specific polymer in a specific application is clearly a complex problem.

Fire-retarding elements

Six elements are particularly associated with fire retardance in polymers, namely, boron, aluminium, phosphorus, antimony, chlorine and bromine. It has already been noted that fire retardants can intervene at one of three points in the burning cycle. However any given fire retardant may act in more than one way so it is not possible to associate an element exclusively with a unique mode of action. Nevertheless phosphorus is strongly associated with changes in thermal degradation processes, chlorine, bromine and antimony with flame quenching and aluminium and boron with inhibition of heat flow, so that consideration of the elements in that order is generally consistent with a clockwise motion round the burning cycle.

6.5.1. Phosphorus compounds

Inorganic phosphates like ammonium phosphate have been used for a very long time for the fire protection of cellulose in the form of wood, paper and cotton. More recently, organic phosphates like trioctyl phosphate and *tris-*(2-chloroethyl)phosphate (III) have come into use and all of these compounds have been found to give a degree of protection to a variety of polymers. In those cases in which the chemistry has been investigated in some detail it seems that the phosphorus-containing salt or ester is acting as a precursor for phosphoric acid and that the fire retardance properties must be associated with the influence of this acid on the thermal degradation reactions which occur in the polymers. This may be illustrated by reference to cellulose and polyurethane.

When cellulose is heated alone to temperatures in excess of 250 °C, about one third of the volatile products consists of water, carbon dioxide, carbon monoxide and acetaldehyde, the remainder being a tarry substance based on the levoglucosan structure, reaction (6.2).

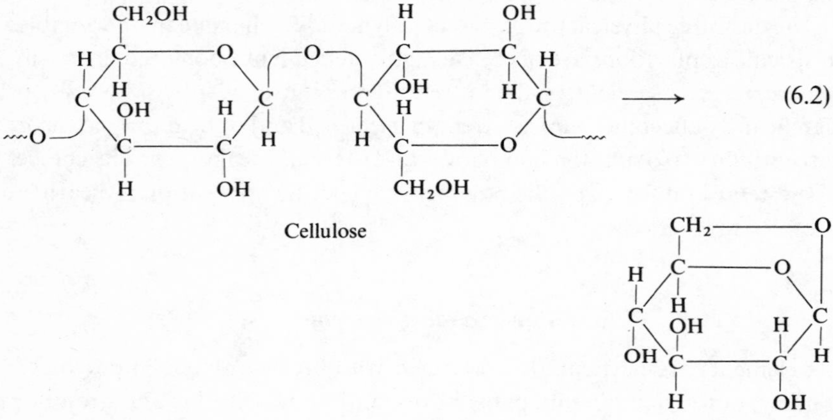

Cellulose

Levoglucosan

$$\hspace{10cm} (6.2)$$

At prolonged heating times or higher temperatures the conversion to these products may be as high as 90%, the residue being a carbonaceous char. In a fire situation the levoglucosan may burn or break down to more volatile, flammable products.

In presence of a phosphate type fire retardant on the other hand, the acid function formed in its decomposition esterifies the hydroxyl groups of the cellulose. The cellulose phosphate then decomposes to produce a a double bond, the acid function being reformed for further reaction. Thus stable conjugated structures are built up in the cellulose and one molecule of phosphoric acid can lead to the production of many double bonds by the sequence shown in scheme 6.2. Conjugated unsaturated structures of this kind are the precursors for the formation of char so that in the presence of phosphates, cellulose forms char and water at the expense of volatile flammable products.

Thus the role of the phosphorus compound in the protection of cellulose is two-fold. First, smaller amounts of flammable products are formed and second, the polymer is protected from the heat of combustion by the layer of char formed on its surface and to a lesser extent by the water evolved.

The mechanism of thermal degradation of a simple polyurethane was

Scheme 6.2 Dehydration of cellulose by phosphoric acid.

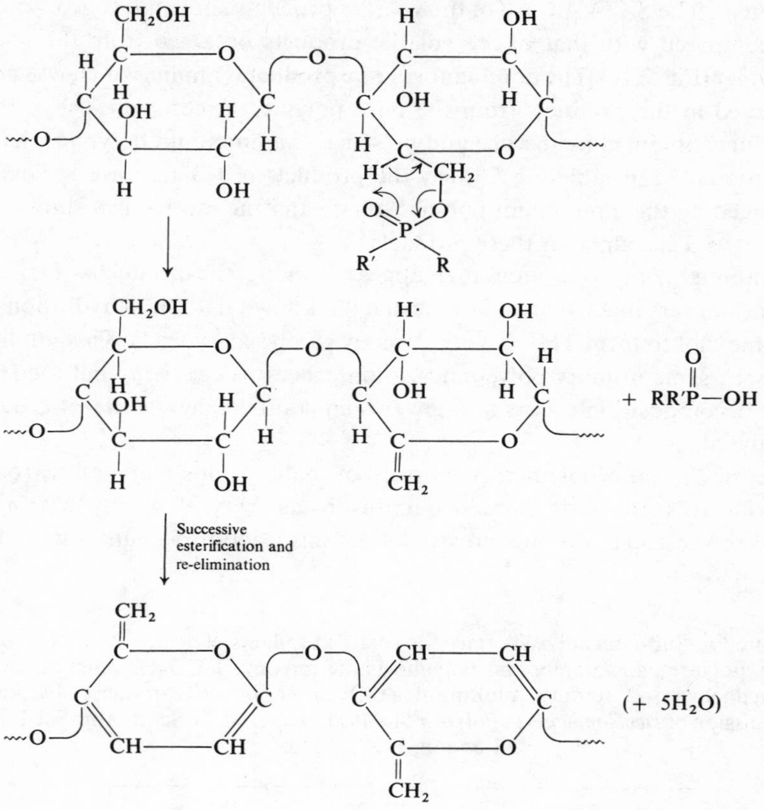

$+ RR'P-OH$

O
‖
$RR'P-OH$

Successive
esterification and
re-elimination

$(+ 5H_2O)$

described in chapter 2. In the presence of ammonium polyphosphate (IV), which eliminates ammonia and water below degradation temperatures to form cross linked polyphosphoric acid (V), reaction (6.3), degradation occurs

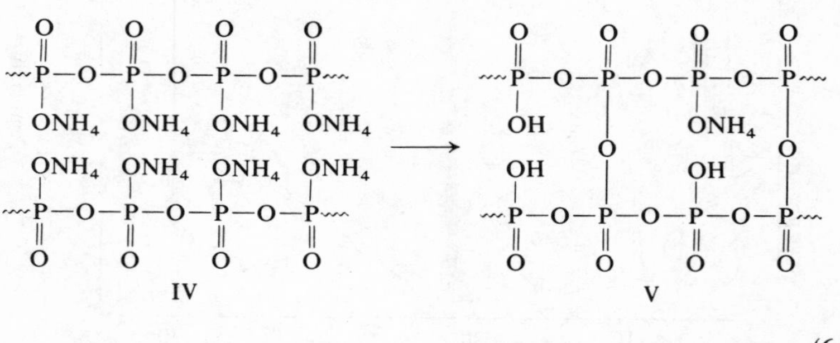

IV   V

$$(6.3)$$

some 30 °C lower and the mechanism of degradation is significantly different. The SATVA trace of the volatile products shown in Fig. 6.1 may be compared with that of the volatile products obtained from the pure polymer (Fig. 2.10). The dominant volatile product is aniline which was not detected in the products from the pure polymer. A complete list of the products obtained in absence and presence of ammonium polyphosphate are presented in table 6.2. Clearly the products of reaction are radically changed by the ammonium polyphosphate and the mechanism shown in scheme 6.3 accounts for these products.

Although only one monomer appears among the products, THF is formed in very much increased yield and it is known that the dehydration of butane diol to form THF is very strongly catalysed by acids. This implies that the same primary polycondensation process is occurring but the fact that decomposition occurs at a lower temperature suggests that it is acid catalysed.

As in the pure polymer, pairs of isocyanate groups eliminate carbon dioxide to form carbodiimide but this reacts very efficiently with the phosphoric acid to produce crosslinks and this reaction accounts for (*a*) the

Figure 6.1. Sub-ambient TVA trace for volatile products of degradation of a 10/1, w/w, polyurethane/ammonium polyphosphate mixture. 1 CO_2; 2 ammonia and formaldehyde; 2 tetrahydrofuran; 4 H_2O; 5 aniline. (Reproduced by kind permission of *Developments in Polymer Stabilisation – 1*, ed. G. Scott, App. Sci. Pub., London, 1979, p. 213.)

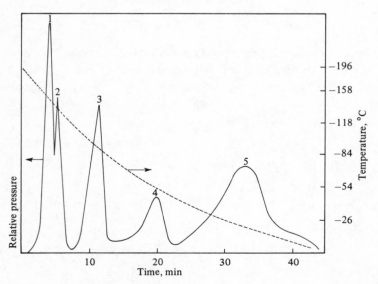

Scheme 6.3 Decomposition of polyurethane in presence of ammonium polyphosphate.

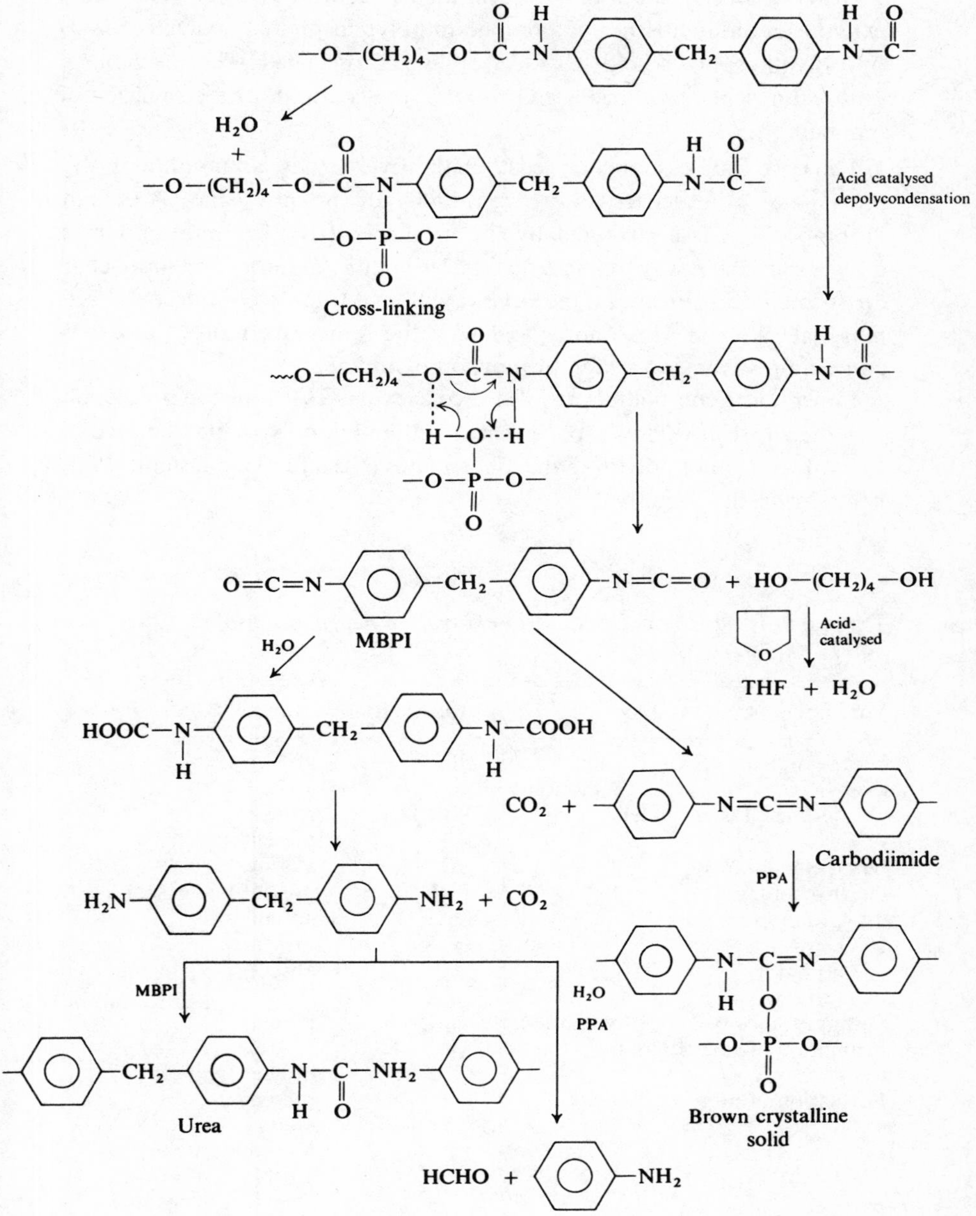

$-O-(CH_2)_4-O-C-N-$⬡$-CH_2-$⬡$-N-C-$

H_2O
+

Acid catalysed
depolycondensation

Cross-linking

$H-O\cdots H$

$-O-P-O-$

$O=C=N-$⬡$-CH_2-$⬡$-N=C=O$ + $HO-(CH_2)_4-OH$

MBPI

H_2O

Acid-catalysed

THF + H_2O

$HOOC-N-$⬡$-CH_2-$⬡$-N-COOH$

CO_2 + ⬡$-N=C=N-$⬡

Carbodiimide

H_2N-⬡$-CH_2-$⬡$-NH_2$ + CO_2

PPA

MBPI

H_2O
PPA

⬡$-N-C=N-$⬡

$-O-P-O-$

⬡$-CH_2-$⬡$-N-C-NH_2-$⬡

Urea

Brown crystalline
solid

$HCHO$ + ⬡$-NH_2$

absence of carbodiimide, (*b*) the larger residue, (*c*) C–O–P and (*d*) C=N structures, as revealed by IR spectral measurements.

Reaction of the isocyanate to form urea structures also occurs to some extent as it did in absence of ammonium polyphosphate but acid catalysis by poly(phosphoric acid) to form aniline is preferred. The mechanism outlined in scheme 6.4 has been proposed to account for the formation of aniline.

Thus we can explain, at least qualitatively, why ammonium poly-phosphate is an effective fire retardant for polyurethane. Although decomposition is accelerated by the presence of the fire retardant it is directed in such a way that less flammable volatile material and more char are formed. In a fire situation the char will tend to protect undecomposed material. We can also understand why the fumes are more obnoxious containing as they do a high proportion of aniline.

Phosphorus compounds are also used as reactives for the fire protection of polymers. For example, phosphorus containing polyols may be used in the preparation of polyurethane, phosphates (VI) and phosphonates (VII) being especially useful.

Table 6.2. *Products and structural change on degradation of polyurethane.*

Pure polyurethane		Abbreviation	Polyurethane + PPA
butane diol		BD	absent
methylene	monomers		
bis-(4-phenyl isocyanate)		MBPI	reduced
CO_2			present
butadiene (minor)		BD	absent
tetrahydrofuran		THF	present (increased)
dihydrofuran (minor)		DHF	absent
H_2O			present
HCN (minor)			absent
CO (minor)		absent	
Disappearance of N—H (urethane link)			present
Formation of carbodiimide			
(—N=C=N—)		CDI	present (reduced)
Formation of urea structures			present
—			aniline
—			formaldehyde
—			C=N structures
—			P—O—C structures
—			large residual char

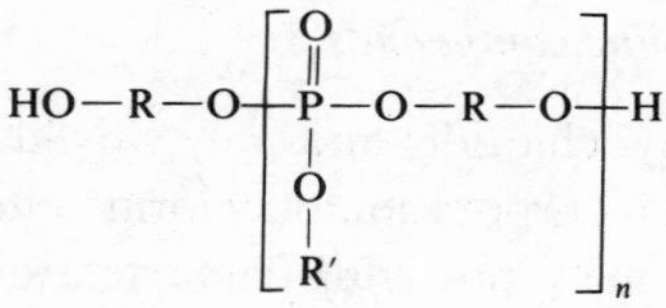

By reactions of such compounds with diisocyanates and various proportions of non-phosphorus-containing polyols, polyurethanes with a range of phosphorus contents may be obtained. (The preparation of polyurethanes is referred to in chapter 2.)

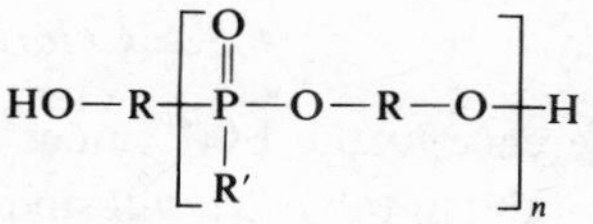

The effectiveness of phosphorus in this form is qualitatively similar to that of a similar concentration of phosphorus atoms in additives and the chemical reactions leading to fire protection are closely comparable. Whether phosphorus compounds are used as additives or reactives will probably depend upon the balance of the usually higher cost of reactives against the advantage of having the fire retarding structures fixed as an integral part of the polymer molecule.

Scheme 6.4 The formation of aniline from polyurethane pyrolysis products in the presence of ammonium polyphosphate.

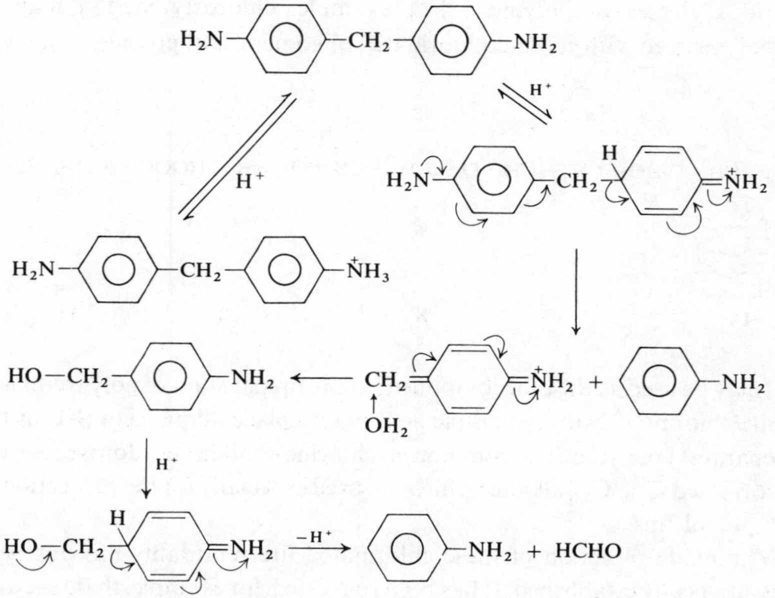

6.5.2. Chlorine and bromine compounds

Reference to the LOI values for poly(vinyl chloride) and poly(vinylidene chloride) in table 2.1 will suggest that the incorporation of chlorine into a polymeric structure must have an extremely powerful flame retarding influence. However the proportion of chlorine is not the only factor since other highly chlorinated polymers, like the polyethers, have very much lower values. Poly(2,2-(dichloromethyl)propylene oxide) (VIII), for example, has an LOI of 23

$$\left[-CH_2-\overset{\displaystyle CH_2-Cl}{\underset{\displaystyle CH_2-Cl}{\overset{\displaystyle |}{\underset{\displaystyle |}{C}}}}-CH_2-O- \right]$$

VIII

which is very much lower than that of poly(vinyl chloride) although somewhat higher than that of the corresponding unchlorinated material (approx. 16–17). In fact the very high LOI values for poly(vinyl chloride) and poly(vinylidene chloride) are attributed principally to the fact that hydrogen chloride is readily lost from adjacent units in the polymer molecule giving a conjugated, cross-linked, carbonaceous char which physically protects the polymer.

A large variety of chlorinated fire retarding reactives have been used for various classes of polymers. For example, chlorostyrene(IX) may be copolymerised with styrene, chlorinated oligomeric compounds of the type

$$CH_2{=}CH \qquad HO-CH_2-\overset{\displaystyle CCl_3}{\overset{\displaystyle |}{CH}}-O-(CH_2-\overset{\displaystyle CCl_3}{\overset{\displaystyle |}{CH}}-O)_n-H \qquad HOOC-(CH_2)_4-COOH$$

IX   X   XI

(X) may be used as the diol component in the preparation of polyurethanes, while chlorine substituted adipic acid may replace adipic acid (XI) in the preparation of nylon. The commonest chlorine containing additives are the chlorinated $C_{10}-C_{30}$ alkanes which are used especially for the protection of the polyolefins.

The mode of action of these chlorinated fire retardants has not been unequivocally established. It has been suggested, for example, that reactives may function by altering the nature of the mechanism and products of degradation or that both reactives and additives may liberate hydrogen chloride which in turn reacts with the polymer. What is quite certain, however, is that with both additives and reactives the chlorine is almost quantitatively converted to hydrogen chloride which finds its way into the flame.

The free radical chain reaction which proceeds in a hydrocarbon flame is complex and involves a large number of primary reaction steps. Two of these primary processes are particularly important in the chain-branching combustion process, namely reaction (6.4)

$$H\cdot + O_2 \rightarrow \cdot OH + \cdot O \cdot \qquad (6.4)$$

and the highly exothermic reaction (6.5)

$$\cdot OH + CO \rightarrow H\cdot + CO_2. \qquad (6.5)$$

The protective function of hydrogen chloride involves its reaction with these highly reactive radical species;

$$H\cdot + HCl \rightarrow H_2 + Cl\cdot \qquad (6.6)$$

$$\cdot OH + HCl \rightarrow H_2O + Cl\cdot. \qquad (6.7)$$

The chlorine atom formed is relatively unreactive and incapable of propagating the oxidation process. It ultimately reacts at a very much slower rate with components of the combustion to regenerate hydrogen chloride. Although the chemistry of this process is still obscure, the reactions could be related to those involved in the termination of alkyl radicals by stable radicals during polymer processing, reaction (6.8)

$$-CH_2\dot{C}HCH_2 - \xrightarrow{Cl\cdot} -CH=CHCH_2 - + HCl. \qquad (6.8)$$

Chlorine-containing compounds are much more effective when they are used in conjunction with certain metal oxides, especially antimony oxide, and this synergistic combination will be discussed in a later section of this chapter.

Bromine is very much more effective than chlorine as a fire-retarding element in both additives and reactives although its compounds are considerably more expensive. Another disadvantage is that the strength of the carbon–bromine compared with the carbon–chlorine bond means that its compounds are much more liable to thermolysis during processing and photolysis during commercial usage, both of which commonly manifest themselves in discolouration of the polymer.

A high proportion of bromine-containing additives are aromatic, like the brominated biphenyls (XII) and biphenyl ethers (XIII), although oligomeric, brominated aliphatic ethers analogous to VIII have also been used.

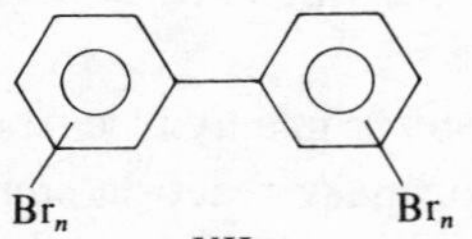

XII 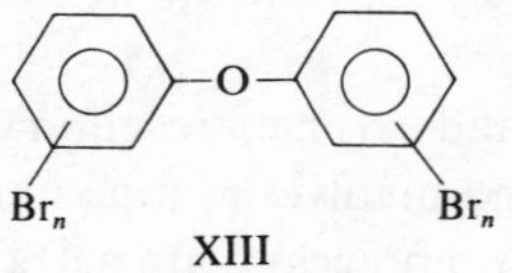   XIII

Bromine-containing monomers have been applied as reactives for a number of different classes of polymers. For example, tetrabromophthalic acid (XIV) and the brominated diol (XV) may partially replace terephthalic acid or ethylene glycol in the preparation of poly(ethylene terephthalate).

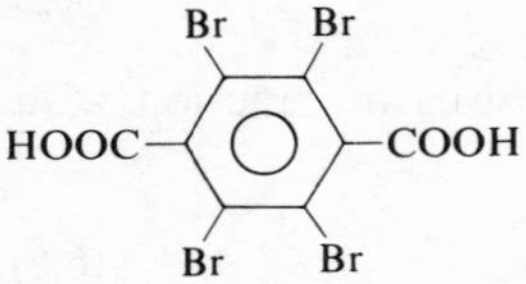

XIV

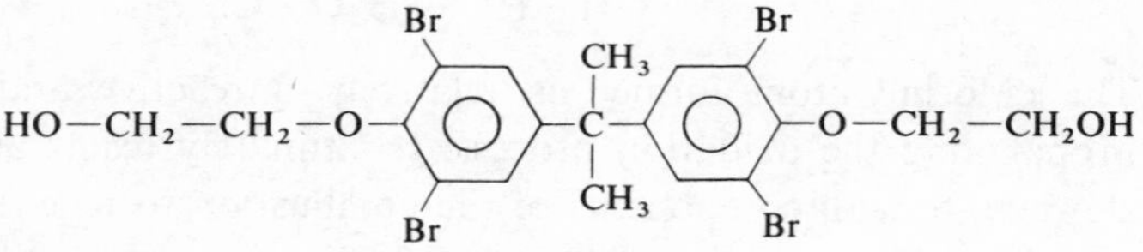

XV

It is vital, of course, that comonomers used in this way should not seriously adversely affect the fibre-forming properties of the polymer. Bromine may also be introduced into acrylic polymers by copolymerisation with brominated acrylic monomers like 2-bromoethyl methacrylate (XVI).

$$
\begin{array}{c}
CH_3 \\
| \\
CH_2{=}C \\
| \\
COOCH_2{-}CH_2Br\cdot
\end{array}
$$

XVI

It is not surprising that copolymers of this monomer with methyl methacrylate and styrene, whose homopolymers give high yields of monomer or monomer related products (dimer, trimer, etc.), yield only the monomers on thermal degradation. On the other hand copolymers with acrylonitrile and methyl acrylate, whose polymers undergo more complicated degradation processes, yield a variety of low molecular weight bromine compounds including hydrogen bromide, methyl bromide, vinyl bromide and dibromoethane.

It seems that the mechanisms of fire retardation by bromine compounds are generally similar to those proposed for chlorine compounds although it seems to be agreed that with bromine compounds retardation is predominantly associated with hydrogen bromide in the gas phase rather than with modifications to the mechanism of degradation in the condensed phase. The hydrogen bromide ultimately formed from all the fire retardants undergoes a series of radical reactions analogous to those set out above for hydrogen chloride, but much more readily due to the greater lability of the hydrogen–bromine compared with the hydrogen–chlorine bond.

Bromine-containing fire retardants are sometimes more effective when small amounts of radical-producing additives like azobisisobutyronitrile or dicumyl peroxide are also present. Originally it was suggested that the free radicals produced by decomposition of these compounds somehow reacted with the polymer to release the bromine atoms more rapidly so that the concentration of hydrogen bromide in the flame was enhanced. More recently it has been shown to be more likely that the free radicals attack the polymer chain, causing chain scission and a rapid decrease in molecular weight. Thus the polymer becomes more mobile so that heat and fuel are lost by the burning polymer dripping away from the main reaction zone. Polystyrene and poly(methyl methacrylate) may be protected in this way although the dripping of burning molten polymer creates new problems of fire spread and hazards to personnel in real fire situations.

6.5.3. Antimony trioxide

Although antimony trioxide (Sb_2O_3 or, more correctly, Sb_4O_6) is very widely used in flame-retarding polymer formulations, and especially with polyolefins, it has little or no fire-retardant activity of its own. It is almost invariably used in conjunction with chlorine or bromine-containing compounds with which it exhibits a synergistic effect. Thus it has been reported that in epoxy resin, 3% of Sb_2O_3 and 5% of Br by weight is as effective as 13–15% of Br.

Since halogen must be present it is reasonable to suppose that an antimony–halogen compound must be formed and that this is probably the oxyhalide (SbOX) which is the product of reaction of Sb_2O_3 and HX, the latter being an ultimate product of decomposition of halogen-containing fire retardants of the type described in the previous section of this chapter.

Of course Sb_2O_3 may be used alone with poly(vinyl chloride) which is its own source of HCl and in this case concentrations of the order of 5–15% by weight of Sb_2O_3 are appropriate.

Antimony oxychloride decomposes into Sb_2O_3 and antimony trichloride ($SbCl_3$), which is volatile, in three fairly discrete steps; reactions (6.9)–(6.11)

$$5SbOCl \xrightarrow{245-280\,°C} Sb_4O_5Cl_2 + SbCl_3 \tag{6.9}$$

$$4Sb_4O_5Cl_2 \xrightarrow{410-475\,°C} 5Sb_3O_4Cl + SbCl_3 \tag{6.10}$$

$$3Sb_3O_4Cl \xrightarrow{475-565\,°C} 4Sb_2O_3 + SbCl_3. \tag{6.11}$$

With both chlorine and bromine compounds the Cl/Sb atomic ratio for optimum fire retardance is close to 3 : 1, which is strong evidence that antimony trichloride is the active agent. Of course $SbCl_3$ cannot be used directly as an additive because of its volatility and hydrolytic instability.

Much of the evidence suggests that antimony/halogen combinations function in the gas phase by reacting directly with the oxidation chain propagating radicals to form hydrogen halide, reactions (6.12)–(6.14)

$$SbX_3 + H \cdot \rightarrow SbX_2 + HX \tag{6.12}$$

$$SbX_2 + H \cdot \rightarrow SbX + HX \tag{6.13}$$

$$SbX + H \cdot \rightarrow Sb + HX. \tag{6.14}$$

The latter destroys additional radicals in reactions 6.6 and 6.7 as discussed in the previous section. It has also been suggested that further reaction of the antimony in the flame results in the formation of relatively unstable monoxide which acts as a catalyst for the recombination of hydrogen, oxygen and hydroxyl radicals.

Thus the primary function of the antimony/halogen combination is to provide radical traps of various kinds in the gas phase. But its efficiency is greatly increased by the fact that the active species are liberated over a wide range of temperatures which match the degradation temperatures of a large number of organic polymers. Thus the 'right place at the right time' requirement is more frequently satisfied.

In addition to these direct gas phase functions, however, fire retardance by the antimony/halogen combination is assisted in at least two other ways. First, reactions (6.9)–(6.11) are all endothermic and this has a cooling effect on the flame. Secondly, with certain polymers, including rubbers, polyesters and polyurethanes, reactions occur in the condensed phase which lead to a more effective protective char than that obtained in absence of antimony.

6.5.4 Aluminium oxide

Historically, aluminium was one of the earliest fire-retarding elements, having been used in the form of alum at least 200 years ago for the protection of cellulose materials like textiles and timber. Alum is still used to some extent in fire extinguishing solutions but it is not a significant fire retardant for commercial polymeric materials. However it is frequently claimed that aluminium, in the form of aluminium oxide trihydrate ($Al_2O_3 . 3H_2O$), is currently the highest-tonnage fire-retarding compound used in polymer technology. Alumina has for long been used as a filler both to modify the physical properties and to reduce the cost of synthetic polymers in certain applications; proportions of up to 70% by weight of alumina may be used, but it is only in the past ten years that the importance of its fire-retardant function has been regarded as being on a par with its function as a filler.

The fact that only the hydrated form of Al_2O_3 is an effective fire retardant provides the essential clue to its action. $Al_2O_3.3H_2O$ comprises 35% by weight of water and dehydrates endothermically in stages between approximately 230 and 330 °C. Thus the fire retardant action clearly involves the following factors. First, the endothermicity of the dehydration process acts as a heat sink to cool the decomposing material. Secondly, the large proportion of water mixed with the volatile flammable products of degradation of the polymer tends to blanket and snuff out the flame. Thirdly, many polymers decompose in the temperature range 230–330 °C so that the 'right place at the right time' principle applies. Fourthly, the involatile ash of Al_2O_3, remaining on the surface of the material after the polymer has decomposed, protects the underlying polymer.

Although these are the most important fire-retarding functions of alumina, there is some evidence of synergism with chlorine-containing compounds. However it is very much less than that of antimony. The mechanism is also obviously different because there is no volatilisation of aluminium chloride into the flame as there is of antimony trichloride. It seems that an aluminium/halogen species is formed but that it acts in the condensed rather than the gas phase.

Alumina is also recognised as an effective smoke suppressant but the chemical mechanisms involved are not clearly understood.

6.5.5. Compounds of boron

Boric acid and borax were probably the most important fire retardant additives for textiles in the nineteenth century. They were used for the protection of cotton at least 150 years ago and even as little as 10–20 years ago they were the most important fire retardants for wool. In these applications it seems that the principal function of these compounds is to form a continuous protective coating over the surface of the fibres. Boric acid decomposes in the following sequence of reactions:

$$2H_3BO_3 \xrightarrow[-2H_2O]{130-180\,°C} 2HBO_2 \xrightarrow[-H_2O]{260\,°C} B_2O_3. \qquad (6.15)$$

Boric acid   Metaboric   Boric oxide
                         acid

The boric oxide softens above 300 °C and flows relatively freely at 500 °C. Borax ($Na_2B_2O_7.10H_2O$), on heating, dissolves in its water of crystallisation which is then driven off leaving an intumescent mass which melts to a clear, fairly mobile liquid.

In practice, mixtures of boric acid and borax are commonly used for at least two reasons. First, the mixture does not crystallise as readily as the separate compounds and thus forms a more continuous and coherent coating on the surface of the fibres. Secondly, the mixture is more effective as a fire retardant because the two compounds have slightly different but complementary functions. Thus boric acid is not so effective as a flame retardant but does inhibit non-flaming surface combustion in the charred material ('afterglow'). On the other hand, borax is a much better fire retardant although very much less effective in extinguishing afterglow. The great disadvantage of these two compounds in many applications is that they are 'non-durable'. That is, being soluble in water they are removed by laundering so that retreatment of the fabric is necessary.

If protection by boric acid and borax was exclusively mechanical then one might expect that they should be effective with a broad range of polymers. This is not so. However in some cases it is clear that the production of char is enhanced and it seems that the explanation is similar to that already advanced above for the mechanism of protection of certain polymers by phosphorus compounds. Either the boric acid esterifies hydroxyl groups, for example in cellulose and phenolics, with subsequent ester decomposition to form unsaturated structures which are the precursors of char, or acid-catalysed reactions occur which drive the thermal decomposition reaction of the polymer in new directions.

Yet another factor in the action of boric acid and borax must be the endothermicity of the dehydration reactions and the blanketing effect of the relatively large proportion of water vapour in the gases feeding the flame.

Other salts of boric acid, and especially zinc salts, have found considerable use as fire retardants for polymers. Their mode of action is quite different however, since, on heating, a thermally stable, liquid surface protectant is not formed. Instead they behave rather like Sb_2O_3 as synergists for chlorine containing compounds. Although their action has not been as thoroughly investigated as that of Sb_2O_3, it is believed that zinc chloride or oxychloride are formed which, being volatile, inhibit oxidation processes in the flame in a fashion closely analogous to $SbCl_3$. Zinc borates are very much cheaper than Sb_2O_3 but they are not as efficient as fire retardants and since, unlike boric acid and borax, they must be incorporated into the polymer during processing, there have been problems associated with the evolution of water of crystallisation at processing temperatures. In the past the zinc borates have usually been complex mixtures of compounds containing zinc oxide and boric oxide in various proportions. However a pure compound with the formula, $2ZnO.3B_2O_3.5H_2O$ has been developed and marketed under the name Firebrake. It is stable to 250 °C which is above the processing temperatures of most common polymers. It is used as a partial replacement for Sb_2O_3 and is effective in poly(vinyl chloride) in concentrations of the order of 5%. In contrast to boric acid and borax these compounds are insoluble and thus durable as fire retardants.

Suggested further reading

1. J. W. Lyons, *The Chemistry and Uses of Fire Retardants*, Wiley & Sons, 1970.
2. C. F. Cullis and M. M. Hirschler, *The Combustion of Organic Polymers*, Oxford, 1981.
3. C. F. Cullis, Metal Compounds as Flame Retardants for Organic Polymers, *Developments in Polymer Degradation* – 3, ed. N. Grassie, App. Sci. Pub., London, 1981.
4. M. Lewin, S. M. Atlas and E. M. Pearce (eds.), *Flame Retardant Polymeric Materials*, Plenum, 1975.
5. W. C. Kuryle and A. J. Papa (eds.), *Flame Retardancy of Polymeric Materials*, Vols. 1–4, Dekker, 1973.
6. M. M. Hirschler, A Study of Flame Retardants, *Developments in Polymer Stabilisation* – 5, ed. G. Scott, App. Sci. Pub., London, 1982, chapter 5.
7. A. Tkac, Flame Retardant Mechanisms: Recent Developments, *Developments in Polymer Stabilisation* – 5, ed. G. Scott, App. Sci. Pub., London, 1982, chapter 6.

7

Degradation in special environments

7.1. Polymers under stress

As synthetic polymers have become established as standard materials of fabrication and construction, it is inevitable that their desirable physical properties have led to a continuous demand for their application in more and more severe environments – at higher temperatures, under high-energy radiation, in chemically or biologically active environments, under extreme mechanical stress, and so on.

In this chapter, brief accounts are given of the more important features of degradation in some of these 'aggressive' environments.

7.2. Degradation in polluted atmospheres

7.2.1. Introduction

Deterioration and ageing of polymers in normal outdoor applications is due essentially to the combined action of sunlight and air and the reactions involved have been discussed in detail in chapters 3 and 4. However experience has shown that other materials which may be present in trace quantities in air may have a significant effect in shortening the useful life of certain polymers. The most important of these are nitrogen dioxide (NO_2), sulphur dioxide (SO_2) and ozone (O_3). All three pollutant gases are present in urban and especially in industrial areas, being associated with the combustion of fossil fuels and with a number of manufacturing processes.

The fundamental reason for the polymer degradant properties of these compounds is easy to discern. Thus NO_2 is an odd-electron molecule and should therefore be capable of initiating free radical reactions. In addition, it absorbs UV light and can dissociate into nitric oxide (NO) and oxygen atoms which are themselves powerful radicals and a precursor of ozone by reaction with the oxygen of the air. Sulphur dioxide is not a free radical and although it strongly absorbs the UV in sunlight it does not dissociate to free radicals. Nevertheless, the activated singlet and triplet states of SO_2 are capable of initiating radical processes. Ozone reacts very rapidly with olefins (see chapter 4), but as will be seen later, it also has the ability to initiate radical reactions.

7.2.2. Nitrogen dioxide

The effect on saturated vinyl polymers of typical pollutant concentrations of NO_2 at ordinary temperatures under UV radiation in air is negligible compared with other photo-oxidative effects which may occur. There is evidence, however, that nitro-groups appear in polyethylene at these temperatures in the presence of high concentrations of NO_2 but this is strictly limited and proportional to the number of abnormal unsaturated structures in the polymer. It has been concluded that addition of NO_2 occurs at the double bonds followed by further addition to the radical to form dinitro- or nitro–nitrite structures.

$$\sim\!CH\!=\!CH\!\sim + NO_2 \longrightarrow \sim\!CH\!-\!\overset{\cdot}{C}H\!\sim$$

$$\begin{array}{ccc}
\sim\!CH\!-\!CH\!\sim & & \sim\!CH\!-\!CH\!\sim \\
|\quad\ | & & |\qquad | \\
NO_2\ \ NO_2 & & NO_2\ \ ONO
\end{array} \qquad (7.1)$$

Unsaturated polymers are much more susceptible to attack by NO_2 than saturated materials. Butyl rubber (polyisobutylene with 0.8 wt per cent of copolymerised isoprene) undergoes chain scission while polybutadiene and polyisoprene rapidly cross link. Addition of NO_2 to the double bonds is obviously the first step followed by addition of a second NO_2 molecule to the resulting radical as for polyethylene (reaction (7.1)). The fact that reaction with the isoprene units in butyl rubber leads to chain scission while polyisoprene and polybutadiene crosslink is perhaps unexpected but must be associated with the nature of subsequent decomposition of the nitrated structures.

At higher temperatures, in the region of 100°C, NO_2 is capable of abstracting hydrogen atoms from the methylene groups in polyethylene and indeed from other saturated vinyl polymers.

$$RH + NO_2 \longrightarrow R\cdot + HNO_2$$
$$R\cdot + NO_2 \longrightarrow RNO_2$$
$$\searrow RONO$$
$$R\cdot + O_2 \longrightarrow RO_2 \qquad (7.2)$$

For example, the tertiary hydrogen atoms in polystyrene are vulnerable and the resultant radicals will add NO_2 or oxygen to form products which at these temperatures may initiate more extensive breakdown of the polymer structure. These high-temperature reactions are not practically important under normal ageing and weathering conditions.

When the hydrocarbon polymer chain is interrupted by amide groups as in the nylons, NO_2 can induce rapid chain scission even at ambient temperatures. The rate and extent of this reaction is strongly influenced by the morphology of the polymer and is strongly inhibited if the amide hydrogen atom is protected by hydrogen bonding, e.g. by benzoic acid.

$$
\begin{array}{c}
\sim\!\!\!\sim C-N\!\!\!\sim\!\!\!\sim \\
\parallel \quad | \\
O \quad H \\
\vdots \quad \vdots \\
H \quad O \\
| \quad \parallel \\
O-C \\
| \\
Ph
\end{array}
$$

In view of the obvious importance of the reactive amide hydrogen atom on the degradation process, the overall reaction has been formulated as shown in reaction (7.3)

$$
\begin{array}{c}
\overset{O \;\; H}{\underset{}{\sim\!\!\!\sim C-N\!\!\!\sim\!\!\!\sim}} + NO_2 \longrightarrow \sim\!\!\!\sim\overset{O}{C}-\dot{N}\!\!\!\sim\!\!\!\sim + HNO_2 \\[2mm]
\swarrow \qquad \searrow \\[2mm]
\overset{O \;\; NO_2}{\underset{}{\sim\!\!\!\sim C-N\!\!\!\sim\!\!\!\sim}} \qquad \text{Chain scission}
\end{array}
\tag{7.3}
$$

Polyurethanes are also degraded by NO_2. Simultaneous chain scission and cross-linking occur although the former predominates. IR absorption bands at 3300, 1695 and $1540\,cm^{-1}$, assigned to absorption by N–H and C–N bonds, decrease in intensity while an absorption at $1140\,cm^{-1}$ develops probably due to incorporation of NO_2 groups into the polymer structure. Carbon dioxide and other unidentified volatile products are formed. There is no doubt that, as in the nylons, the reaction is initiated by abstraction of the reactive N–H hydrogen atom by NO_2. The radicals thus formed either react to form cross-links or undergo scission followed by elimination of CO_2, the precise reaction mechanism being unclear.

7.2.3. *Sulphur dioxide*

Sulphur dioxide is less reactive than NO_2 towards organic molecules principally because of the radical nature of the latter. However the ability of SO_2 to absorb UV radiation forming excited singlet and triplet states makes it a much more potent degradant for polymers under normal weathering conditions. The excited SO_2 molecules can react with oxygen and moisture in the atmosphere to form both ozone and sulphuric acid

$$SO_2(\text{singlet}) + O_2 \xrightarrow{h\nu} SO_4 \qquad (7.4)$$

$$SO_4 + O_2 \rightarrow SO_3 + O_3 \qquad (7.5)$$

$$SO_3 + H_2O \rightarrow H_2SO_4 \qquad (7.6)$$

which are themselves powerful degradants, the former for rubbers and the latter for the polyesters which are susceptible to acid hydrolysis. Triplet SO_2 may also react directly with hydrocarbon molecules to form sulphinic acids,

$$SO_2{}^* + RH \longrightarrow R-S\underset{O}{\overset{OH}{\diagup\diagdown}} \qquad (7.7)$$

or abstract hydrogen atoms or react with olefinic structures to initiate radical processes (reactions 7.8 and 7.9) including oxidation by subsequent reaction with ground state oxygen.

$$SO_2{}^* + RH \longrightarrow R\cdot + H\dot{S}O_2 \qquad (7.8)$$

$$SO_2{}^* + \overset{}{\underset{}{\diagup}}C=C\overset{}{\underset{}{\diagdown}} \longrightarrow \overset{}{\underset{}{\diagup}}C-C\overset{}{\underset{}{\diagdown}}_{\underset{O \quad O\cdot}{\overset{|}{S}}} \qquad (7.9)$$

It was seen in an earlier section (chapter 5, section 5) that sulphur acids are important peroxidolytic antioxidants in the absence of light. In general, however, their effect in the outdoor environment is deleterious due to their photo-excitation to chemically reactive species. Most common saturated polymers undergo some degree of chain scission on exposure to the potent combination of sulphur dioxide, oxygen and UV irradiation, but poly-ethylene and to a lesser extent polypropylene undergo rapid cross-linking. Nylon is much less susceptible to attack by SO_2 than by NO_2, the precise mechanism of chain scission and cross-linking by SO_2 is not clearly understood.

7.2.4. Ozone

Molecular oxygen undergoes photolysis in the stratosphere to give oxygen atoms which in the presence of an inert body (M), react with more oxygen to give ozone, reactions (7.10) and (7.11).

$$O_2 \xrightarrow[\text{($\lambda < 242$ nm)}]{h\nu} 2O^{\bullet} \tag{7.10}$$

$$O^{\bullet} + O_2 \xrightarrow{M} O_3 \tag{7.11}$$

Ozone is also formed in the troposphere (lower atmosphere), but in this case, since the shorter wavelengths of the Sun's spectrum are absorbed by the upper atmosphere, another source of energy has to be provided. Peroxyl radicals, which are intermediates in the autoxidation of volatile hydro-carbons in industrial atmospheres, oxidise molecular oxygen to ozone in the presence of nitric oxide which acts as an oxygen transfer catalyst, reactions (7.12)–(7.14)

$$RH \xrightarrow{RO^{\bullet}/O_2} ROO^{\bullet} + ROH \tag{7.12}$$

$$ROO^{\bullet} + NO \longrightarrow RO^{\bullet} + NO_2 \tag{7.13}$$

$$NO_2 + O_2 \xrightarrow{h\nu} O_3 + NO. \tag{7.14}$$

The polluted air is transported by convection currents to the Earth's surface and in some major industrial conurbations (e.g. Los Angeles) the ozone concentration may reach levels which are dangerous to human beings as well as being damaging to polymers.

Ozone is rapidly de-activated by contact with organic materials and the particularly facile and damaging addition of ozone to the double bonds in rubber has already been described in chapter 4 (section 4.2.5). However the reaction of ozone is not limited to unsaturated polymers. It also reacts, albeit more slowly, with polymers containing labile hydrogens and this leads to the formation of free radicals which are the cause of subsequent thermal-oxidative chain reactions at relatively low temperatures. The formation of carbonyl compounds in polymers such as polystyrene and the polyolefins takes place rapidly when they are subjected to ozonised air. Although the chemistry has not been fully elucidated, the first stage of the reaction appears to involve the abstraction of the most labile hydrogen, even in the case of polystyrene which contains aromatic double bonds, scheme 7.1.

Scheme 7.1 Reactions of polystyrene with ozone.

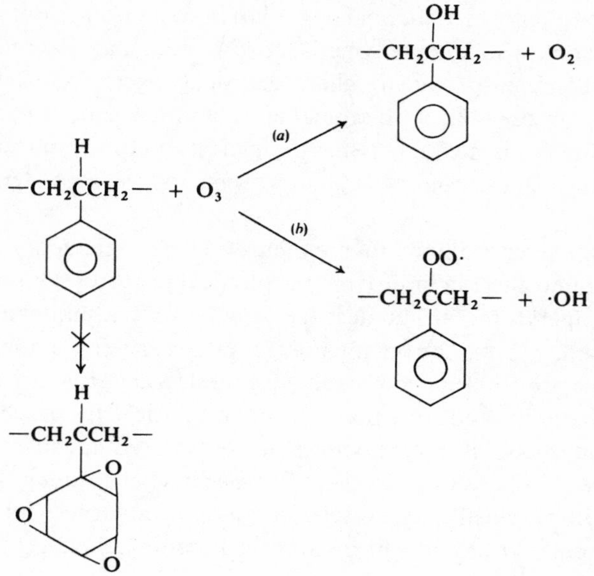

Since ozone normally reacts with polymers at room temperature and hence the rate of radical formation is high, alkylperoxyl termination reactions are more important than they are in thermal oxidation. Consequently about 85% of the products are formed by rearrangement and side reactions of the intermediate species. There is no doubt however that reaction (b) in scheme 7.1 is an important initiating process for many polymers subjected to polluted atmospheres.

7.3 Degradation at high temperatures

7.3.1. *High-temperature polymers*

The problems associated with the application of polymers at elevated temperatures result primarily from instability to thermal decomposition and susceptibility to chemical attack, especially by oxygen. The resulting physical and chemical changes lead to the loss of useful mechanical properties.

There are a number of examples in chapter 2 which illustrate how the thermal stability of linear-chain aliphatic polymers may be improved somewhat by interference with the degradation mechanism. For example, the stability of polyacetals and poly(methyl methacrylate) may be improved marginally by elimination of the terminal hydroxyl or of unsaturated structures through which thermal degradation to monomer is initiated, or in the latter case also by copolymerising with methyl acrylate in order to block the monomer-producing chain depropagation process. But these approaches can only have a marginal effect and it has been necessary to consider more fundamental polymer design features at the molecular level in order to achieve significant improvements in high temperature performance.

An increase in crystallinity due to increased chain interaction can often have a profound effect in improving the physical properties of polymers at elevated temperatures, although it has usually only a minimal effect on thermal stability. Thus the possibilities for closer packing of the chains in high-density polyethylene compared with the low-density variety lead to higher crystallinity and an enhancement of physical properties such as stiffness and modulus which allows it to be used at slightly higher temperatures. The effect is even more obvious by comparing polyethylene with the nylons which may be regarded as polyethylene with regularly distributed polar groups at which strong chain interaction occurs through hydrogen bonding.

$$\sim\!CH_2\!-\!CH_2\!-\!\overset{\parallel}{\underset{O}{C}}\!-\!\overset{\mid}{\underset{H}{N}}\!-\!CH_2\!-\!CH_2\!\sim$$

$$\sim\!CH_2\!-\!CH_2\!-\!\overset{\mid}{\underset{}{N}}\!-\!\overset{\parallel}{\underset{}{C}}\!-\!CH_2\!-\!CH_2\!\sim$$

The enhanced physical properties of nylon 6,6 and its application as an engineering material at much higher temperatures than polyethylene are a reflection of its higher melting and glass transition temperatures (T_m and T_g are respectively 540 K and 333 K for nylon and 410 K and 188 K for polyethylene).

Considerable improvements in performance at elevated temperatures may be achieved by the incorporation of aromatic structures into the polymer chain back-bone. Not only does the chain stiffening effect of these structures improve the mechanical properties, but the replacement of aliphatic by aromatic hydrogen atoms tends to reduce chemical activity. On this basis polyphenylene (I) should be expected to have optimum stability.

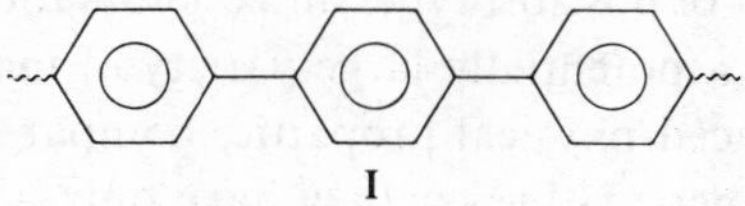

I

It is in fact stable to at least 500 °C but the extreme chain stiffness results in an insoluble, infusible and generally quite intractable material which has been metaphorically described as 'brickdust'. An improvement in physical properties at the expense of stability may be achieved by the insertion of inter-aromatic linking groups of a variety of types and, if we describe the linear aliphatic polymers like poly(vinyl chloride), poly(methyl methacrylate) and polypropylene as the first generation of general-purpose commercial polymers, then it is significant that the second generation of engineering polymers are predominantly of the main chain aromatic group type. This includes such well-established commercial materials as poly(ethylene terephthalate) (II), polycarbonates (III), epoxy resins (IV), polyurethanes (V) and the so-called 'aromatic nylon' (VI).

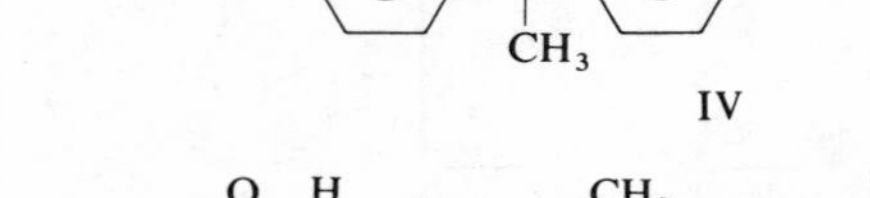

II

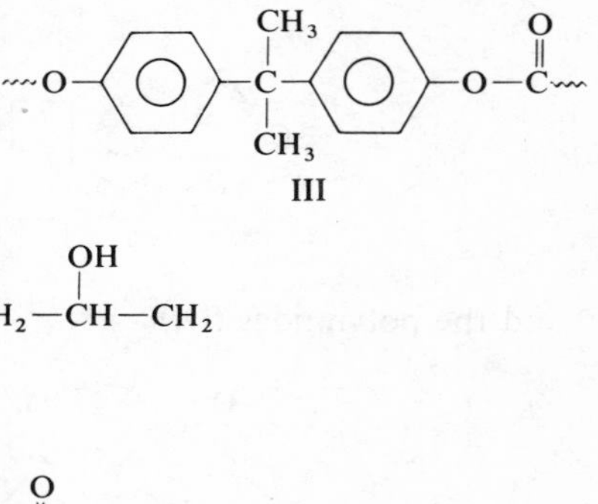

III

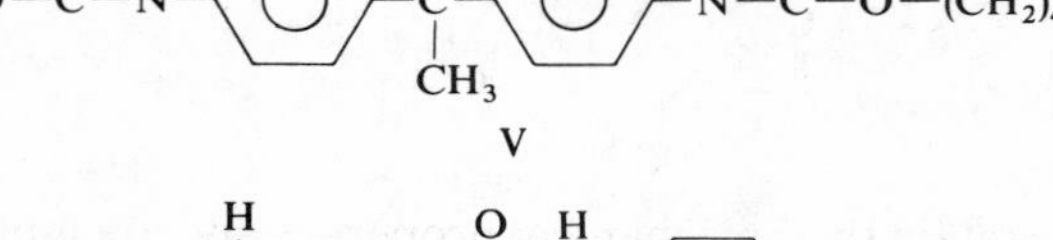

IV

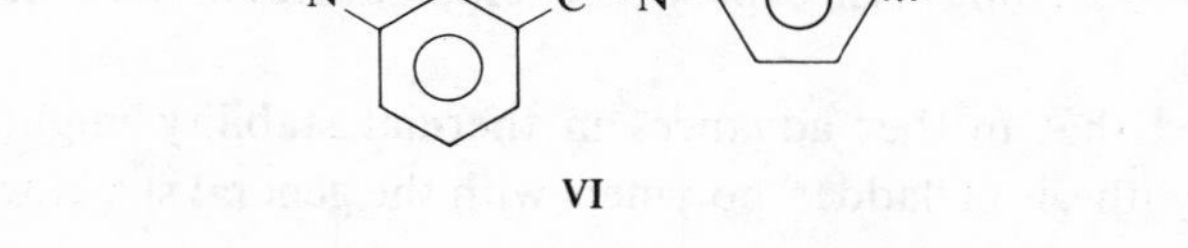

V

VI

Polymers III, IV and V are represented as being derived from 4,4'-isopropylidenediphenol (commercially described as bisphenol A) but any one of a large number of diphenols may be used in their preparation. Thus the very large number of possible types and combinations of interaromatic structures represents a potentially large variety of materials with a wide range of much enhanced physical properties compared with those of the first generation polymers. However they have only a small advantage in thermal stability which limits their long term use to temperatures below 250 °C.

The limited stability of the aromatic polymers is associated with the inter-aromatic groups and further progress towards high temperature materials has depended upon their elimination from the polymer structure. A large number of polymers of this kind have been synthesised and several are already being manufactured commercially. These include the benzimidazoles (VII), benzoxazoles (VIII)

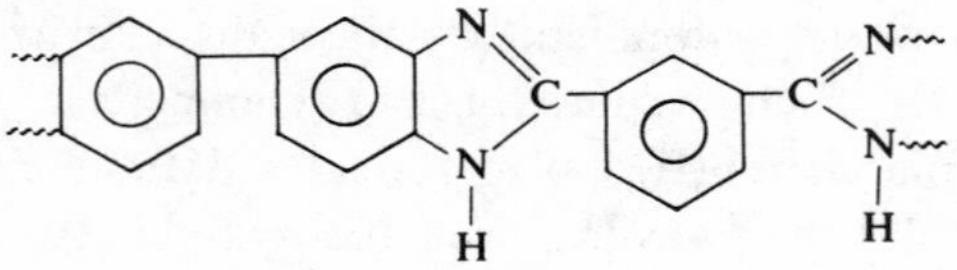

VII

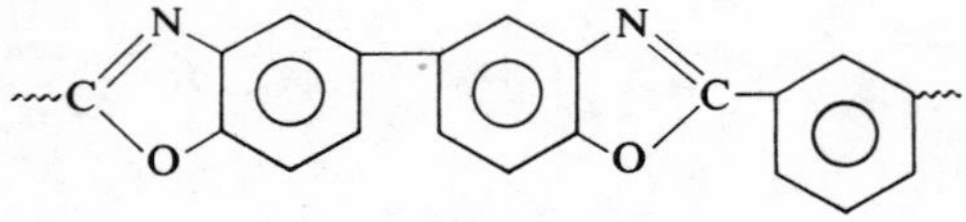

VIII

and the polyimides (IX)

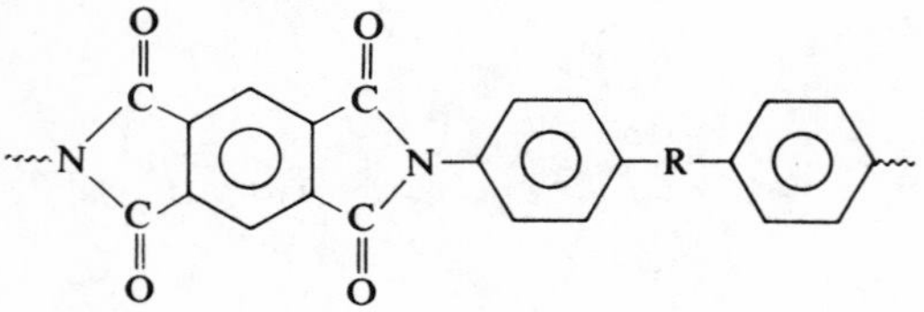

IX

All of these materials retain their mechanical properties and are immune to thermal degradation on long-term exposure to temperatures of the order of 500 °C.

It was reasoned that further advances in thermal stability might be achieved by the synthesis of 'ladder' polymers with the general structure X

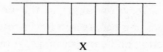

X

Being linear, the polymer should be tractable, but being two-stranded with regular cross-links between the strands, such a polymer should be highly stable because chain scission would require the breaking of at least two bonds within a small cycle of atoms – a rare event if bond scission occurred at random throughout the structure.

Polybenzimidazopyrrolones (XI) and polyquinoxalines (XII) are two of a number of such ladder polymers which have been prepared but seem to have little or no advantage in stability over materials of types VII, VIII and IX.

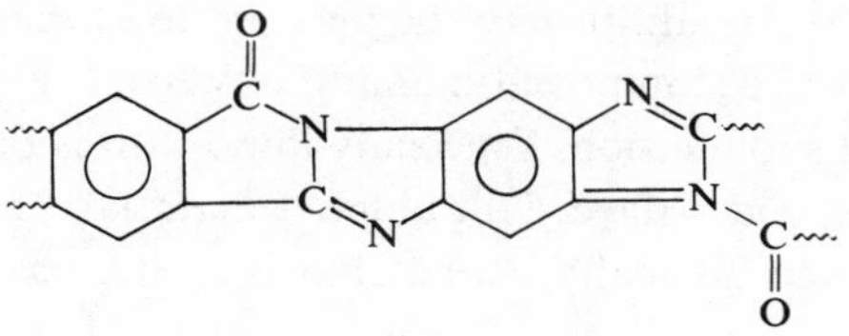

XI

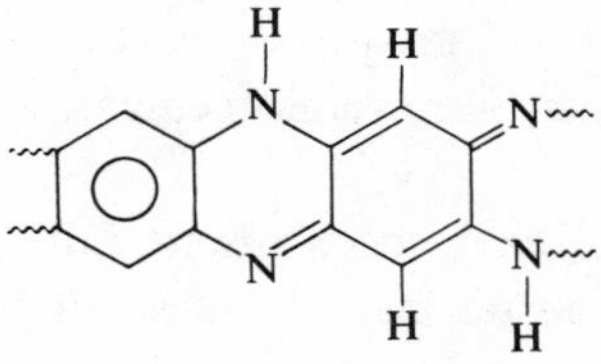

XII

This may be due to the difficulty of synthesising perfect ladder structures or to the fact that bond scission within a small cycle of atoms triggers off more extensive breakdown of the cycle in which it occurs so that chain breaking follows each primary bond scission.

7.3.2. Ablation

Heat resistance of a much higher order of magnitude was demanded by the re-entry problem associated with space flight. Even the most refractory materials re-entering the Earth's atmosphere at high velocity from space will be vapourised by the intense heat liberated as the kinetic energy of the material is converted to heat as it is slowed down by friction with the Earth's atmosphere. This may be alleviated to some extent by arranging that re-entry occurs at a very low angle to the Earth's surface so that deceleration is less and heating occurs at a lower rate. However opposing cooling effects by conduction and convection are minimal in rarified air at high altitudes so that the basic problem remains. An additional requirement of materials used for these purposes is that they should have low density which in any case rules out most of the traditional refractory materials. This problem, at first sight apparently insoluble, has been very satisfactorily solved by applying the principle of ablation or burning away. In space technology this principle has also been used to protect those parts of rocket propulsion systems which are subjected to the very high temperatures (up to 4000 K) associated with fuel combustion.

The ablative materials from which the re-entry and rocket motor heat shields are fabricated are designed for maximum absorption of heat during the ablation process. In addition to the heating up of the material, heat is absorbed in chemical decomposition and in physical changes like melting, vapourisation and sublimation. Especially important is the loss of radiant energy from the charred surface of the ablative material. Thus the formation of a compact and mechanically strong char is a vital part of the ablation process.

In order to maximise these modes of heat dissipation, ablative materials are formed as composites or mixtures, usually with four principal components, namely polymer, fibre reinforcement, low-density filler and subliming additive.

A number of polymers have been used, but principally phenol-formaldehyde resins, epoxy resins and silicones. All of these materials are cross-linked and decompose on heating to give a char with suitable properties. From the first two the char is carbonaceous and from the third siliceous (SiO_2).

The fibre re-inforcement is designed to strengthen the char and more specifically to reinforce the cracks which inevitably occur during the degradation process. Silica fibres are most commonly used and conveniently form a viscous liquid surface layer at very high temperatures. Carbon fibres have a weight advantage but are susceptible to oxidation.

Low-density fillers are used to reduce the thermal conductivity of the material in order to protect the underlying surface more effectively. These are commonly powdered cork or silica microspheres which will clearly also contribute to the carbonaceous or siliceous surface layer later in the ablative process. The subliming additive is commonly a polymer such as nylon 6,6 which melts and readily degrades at relatively low temperatures to give a large volume of gaseous products. These processes have a cooling effect but the escape of the gaseous products of decomposition assists in the formation of a char of appropriate porosity.

Clearly a large number of complex factors are involved in the formulation of an ablative material appropriate to a specific application.

7.4. Mechanical and ultrasonic degradation

7.4.1. Introduction

Polymer degradation may be induced by purely mechanical forces. This usually manifests itself by a decrease in molecular weight. In bulk polymers it may occur, for example, during processing, grinding, milling or sawing, or when the polymer is subjected to stress or shear. Practical use is made of this kind of degradation in the 'mastication' of rubber to reduce its viscosity and thereby make it more readily processable (see section 4.2.1). Mechanical degradation also occurs in solution as a result of shaking, beating, high-speed stirring, droplet formation or turbulent flow. It is this kind of degradation which is at least partly responsible for the deterioration of 'viscostatic' or 'multigrade' lubricating oils in which the viscosity of a light oil has been artificially increased by addition of an appropriate polymer. Similarly the degradation of polymers during screw extrusion is at least partly mechano-chemical (see section 4.2.1(*b*), (*c*)).

A quantitative analysis of mechanical degradation is also relevant to the discovery that certain polymers present in concentrations of a few parts per million (ppm) can reduce the frictional drag between a flowing liquid and a solid/liquid interface. This phenomenon has possible application in, for example, reducing the frictional drag in liquids flowing through pipes and in ships at sea. The applications have been limited, however, by problems of polymer degradation due to turbulent flow, so that the drag reducing effect rapidly decreases.

Mechanical degradation in solution has usually been measured by the decrease in the viscosity of the solution, but care must be observed in the interpretation of experimental results since viscosity changes can sometimes be the result of the dispersion of molecular aggregates. Such a thixotropic effect is usually reversible, however, in contrast to true degradation in which permanent bond scission occurs. For these reasons, extreme care must be exercised in the handling of polymer solutions when molecular dimensions are to be measured by viscometric techniques.

Reduction in molecular weight may also be induced in solution by ultrasonic irradiation. The characteristics of the reaction are very similar to those of other kinds of mechanical degradation so that it may be considered as a special case of this kind of breakdown.

7.4.2. Quantitative aspects of ultrasonic degradation

It has been rather difficult to put mechano-chemical reactions on a quantitative basis because of the difficulty of defining the stresses placed on the chemical bonds in the polymer in terms of the primary applied mechanical energy. This has also made it rather difficult to compare results obtained by different investigators. The view has been expressed that mechano-chemical reactions may really be thermal reactions occurring in 'hot-spots' within the polymer as a result of the conversion of mechanical to thermal energy. This heat would be rather inefficiently dissipated in polymers which usually have low thermal conductivities. These possibilities have been eliminated however by work at low temperatures and relatively low rates of shear.

Of all kinds of mechanical degradation, ultrasonic degradation is probably most amenable to quantitative treatment in the sense that the intensity and frequency of the radiation can be measured much more readily than shearing or impact forces. Thus, a great deal more may be learned by expressing the extent of the reaction in terms of acoustic energy input rather than time. Acoustic energy is defined by

$$Q = ISt/vc \qquad (i)$$

in which I is the intensity of ultrasonic vibration in watts cm^{-2}, S is the area of the bottom of the vessel and T, v and c are respectively time of exposure, and the volume and concentration of the material. Thus Q is a measure of the amount of energy per unit mass to which the polymer has been exposed. Figures 7.1 and 7.2 illustrate the kind of additional information which the use of Q values can give. The latter figure shows quite clearly, while the former does not, that in the range of intensities from 5–30 watts cm^{-2} equal amounts of degradation occur for equal expenditure of energy. On the other hand, increasing the intensity beyond 30 watts cm^{-2} results in more efficient breakdown.

7.4.3. Mechanism of bond scission

Scission of polymer molecules by mechanical forces must be associated with the interactions between these molecules. In solid polymers, in which individual molecules are intimately entangled and relative movement is severely restricted, mechanical stress will lead to scission if it is applied at a rate greater than that at which relative movement of molecules can occur and to an extent greater than can be accommodated by the distortion of primary valency bonds. In solution, during turbulent flow or droplet formation, entanglements and interactions of chains will be important, but the stresses which lead to scission are probably associated with the violent movement of solvent molecules relative to the less mobile polymer molecules. During ultrasonic degradation, the same kind of differential movement of solvent and polymer molecules must occur. This movement of solvent relative to polymer molecules is associated with the formation, vibration and collapse of cavities within the solution. The kind of cavitation with which ultrasonic degradation is associated is due to the liberation of tiny bubbles of dissolved gas which vibrate and ultimately collapse with considerable violence. The small solvent molecules can react to this violent movement very much more readily than the macromolecules and the differential movement of the two places considerable strain on the bonds of the latter.

Figure 7.1. Change of chain length with time during degradation of polystyrene in benzene solution by various intensities (watt cm^{-2}) of ultrasonic radiation. $\bigcirc$ 5; $\square$ 10; $\triangle$ 15; $\bullet$ 30; $\triangle$ 50; $\blacksquare$ 100; $\blacktriangle$ 150; $\blacktriangle$ 250. (Reproduced by kind permission of *Polymer Science*, ed. A. D. Jenkins, North Holland Publishing Company, 1972, p. 1507.)

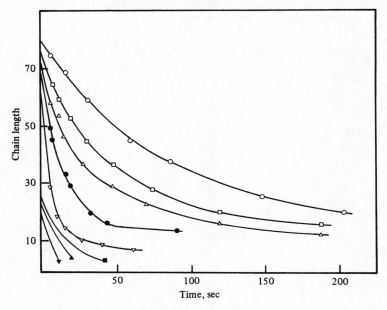

In the light of these mechanisms of bond scission under mechanical stress it is not difficult to understand the influence of certain molecular structural features on the tendency of polymers to undergo mechanical degradation. Thus more rigid molecules are more easily ruptured because the movement of the molecules under stress will be inhibited. On the other hand, the greater the number of chains per unit volume the less will be the stress per chain. For example, the rate constant for bond scission in polystyrene is approximately four times that in polyvinyl alcohol; the chain density of the latter is about four times the chain density of the former. Chain branches, at least in the case of starch, are preferentially ruptured while more extended molecules are broken down more rapidly than coiled molecules. Thus comparison of the two polypeptides, gelatin and casein shows that the molecular weight of more extended gelatin molecules decreases much more rapidly than that of the coiled casein. It is not surprising either that the rate of bond scission increases with molecular weight.

7.4.4. *Quantitative aspects of changes in molecular weight*

In all forms of mechanical degradation there appears to be an ultimate value of molecular weight to which each polymer tends under a given set of

Figure 7.2. Change of chain length with Q value during degradation of polystyrene in benzene solution by various intensities (watt cm^{-2}) of ultrasonic radiation. Intensities as in Fig. 7.1. (Reproduced by kind permission of *Polymer Science*, ed. A. D. Jenkins, North Holland Publishing Company, 1972, p. 1508.)

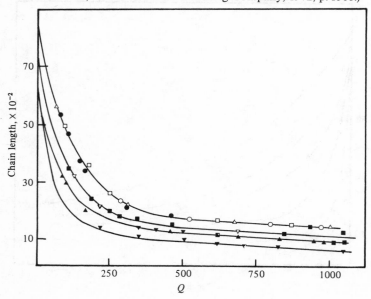

conditions. It has been shown, for example, that under fixed conditions of ultrasonic irradiation the molecular weight of three fractions (initially 850 000, 350 000 and 195 000) of polystyrene in toluene all decreased to 30 000. It also appears to be quite generally true that whether the starting material is a sharp fraction or a broad distribution the ultimate product of long exposure exhibits a narrow distribution. On this evidence it is quite clear that any theory of mechanical degradation must assume that large molecules are much more susceptible to scission than small molecules. This seems to be due to the fact that under conditions of shear degradation, tensile forces on individual chain bonds become very much smaller as the chain ends are approached. The ultimate limiting molecular weight is thus due to the fact that the tensile forces have become too small for scission to occur because the small molecules can adjust themselves to the stress by moving with respect to one another and the surrounding solvent molecules.

The existence of a limiting molecular weight is accounted for by the equation

$$\mathrm{d}x/\mathrm{d}t = k(P_t - P_{\mathrm{lim}}) \qquad\qquad (ii)$$

which states that the rate of bond scission ($\mathrm{d}x/\mathrm{d}t$) is proportional to the difference between the molecular weight at time t (P_t) and the limiting molecular weight (P_{lim}) after prolonged exposure to mechanical stress. This relationship does account qualitatively for the behaviour of a number of systems, for example, the ultrasonic degradation of poly(methyl methacrylate), poly(vinyl acetate) and polystyrene in benzene solution.

The relationship in equation (*ii*) implies that all bonds within a certain distance from chain ends are not ruptured while all others are equally readily broken. It is to be intuitively anticipated, however, that there will be a more gradual increase in the rate constant for bond scission from zero to a maximum value on moving from the ends to the centre of a large molecule.

The detailed analysis of such a system has become possible by the mathematical treatment of gel permeation chromatography (GPC) data. The series of GPC curves illustrated in Fig. 7.3 were obtained at intervals in the course of the degradation of poly(α-methyl styrene) in toluene solution during repeated flow through a capillary. Analysis of curves of this type for a series of starting molecular weights leads to families of curves of the types illustrated in Fig. 7.4. Each curve in Fig. 7.4 represents a different molecular weight and relates the rate constant for bond scission, k, to the distance, I, of that bond from the middle of the polymer molecule. Thus bonds are seen to be more susceptible to mechanical degradation the nearer they are to the middle of the polymer molecule and the effect obviously increases with molecular weight.

7.5. Degradation by high-energy radiation

7.5.1. Introduction

It was pointed out in chapter 3 that the photons of visible and UV radiation, which initiate the photodegradation of polymers, are associated with energies which are of the order of magnitude of bond dissociation energies. The description 'high-energy radiation' implies forms of radiation the energy of whose photons is greater than this by a factor of 10^3 to 10^6. γ-rays have been commonly used in polymer investigations and ^{60}Co and ^{137}Co which emit photons with energies of approximately 1 MeV ($= 10^6$ eV) have been convenient sources. By comparison, the photons of UV radiation have energies of the order of 4–5 eV (~ 400 kJ). Fast electrons generated in van der Graaf generators and linear accelerators have also been used in polymer investigations but β-rays, x-rays, α-particles and neutrons are all forms of radiation with comparable quantum energies.

The most obvious difference between photo- and high-energy radiation in the initiation of chemical changes resides in the extreme selectivity of the former. Thus photons of visible and UV radiation are absorbed only at

Figure 7.3. Change in molecular weight distribution of poly(α-methyl styrene) in toluene solution due to degradation during flow through a capillary. Numbers on curves refer to increasing reaction times. (Reproduced by kind permission of *Degradation and Stabilisation of Polymers*, ed. G. Geuskens, App. Sci. Pub., London, 1975, p. 20.)

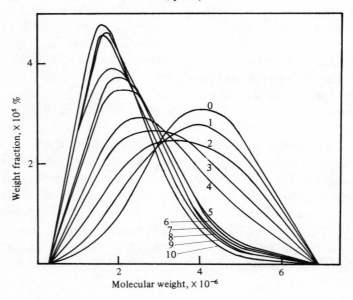

appropriate chromophoric groups. The energy is redistributed according to well defined rules and the bonds finally broken are usually in the close vicinity of these sites of absorption.

High-energy radiation is much less selective. The most significant effect of most forms of high-energy radiation on matter results from interaction with the orbital electrons in a more or less random fashion. In losing its energy in a succession of such interactions, each photon excites a number of electrons as in reaction 1 of scheme 7.2. If the excitation energy is less than the ionisation energy, the excited molecule may dissociate into radicals (reaction 2), otherwise ionisation will occur (reaction 3). The positive polymer ion radicals may subsequently react with a free electron to reform an excited molecule (reaction 4) or they may decompose to form a radical and a positive ion (reaction 5). The electrons may react with polymer molecules to form negative ion radicals (reaction 6) which decompose to form a radical and a negative ion (reaction 7). The presence of radicals in irradiated systems is easily detected by electron spin resonance spectroscopy and ions by spectroscopic measurements.

Figure 7.4. Rate constants k for mechanical scission of poly(α-methyl styrene) as a function of distance I of the bond broken from the centre of the molecule. The numbers on the curves are a function of molecular weight. (Reproduced by kind permission of *Degradation and Stabilisation of Polymers*, ed. G. Geuskens, App. Sci. Pub., 1975, p. 21.)

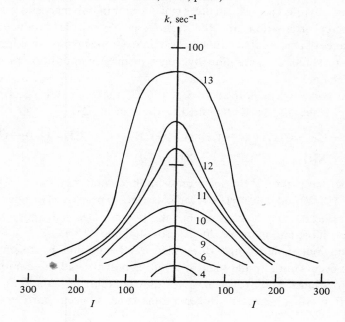

Scheme 7.2 Mechanism of degradation by ionising radiation.

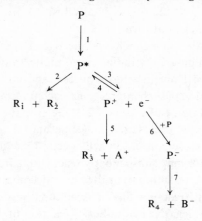

The overall picture is very much more complicated than is represented in scheme 7.2. For example, charge and energy transfer processes occur and ions are destroyed in ion–ion and ion–electron interactions but the most significant and obvious effect of high-energy irradiation of polymers is to induce radical reactions.

7.5.2. Chemical changes in polymers

The most obvious chemical changes which occur in polymers as a result of high-energy irradiation are the formation of volatile products, the appearance of unsaturated structures, chain scission and cross-linking. The volatile products are commonly hydrogen or the decomposition products of side chains.

When polyethylene is irradiated at 196 °C, a six-line ESR spectrum is observed which has been attributed to the radical XIII

$$\text{\textasciitilde}CH_2{-}\dot{C}H{-}CH_2\text{\textasciitilde} \qquad \text{\textasciitilde}CH{=}CH{-}\dot{C}H\text{\textasciitilde} \qquad \text{\textasciitilde}(CH{=}CH)_n{-}\dot{C}H\text{\textasciitilde}$$
$$\text{XIII} \qquad\qquad\qquad \text{XIV} \qquad\qquad\qquad \text{XV}$$

At higher temperatures the spectrum is more complicated due to the fact that allyl (XIV) and polyenyl (XV) radicals are also present. Hydrogen is the only significant volatile product and the polymer becomes progressively more insoluble and infusible due to cross-linking.

The obvious explanation of this evidence is that carbon–hydrogen bond scission occurs initially and that the resulting polymer radicals combine to form cross links. The hydrogen atoms may combine in pairs but a large proportion will abstract hydrogen atoms from adjacent carbon atoms leading to unsaturation. Thus the overall process may be represented by reaction 7.15

$$\sim\!CH_2\!-\!CH_2\!\sim \xrightarrow{\gamma\ rays} \sim\!CH_2\!-\!\dot{C}H\!\sim\ +\ H\cdot \longrightarrow \sim\!CH\!=\!CH\!\sim\ +\ H_2$$

$$\sim\!CH_2\!-\!CH\!\sim$$
$$\underset{\displaystyle \sim\!CH_2\!-\!CH\!\sim}{|}$$

$$H_2 \tag{7.15}$$

The combination of polymer radicals to form cross links is identical to that which occurs during photo-degradation of polyethylene (chapter 3) and it is perhaps surprising that there is less evidence of chain scission of the polyethylene radicals under high energy radiation than there is under UV radiation. The explanation is perhaps that the nature of the initiation process under high-energy radiation is such that polymer radicals are formed in high concentrations in close proximity to one another so that second-order cross-linking reactions are favoured compared with first-order chain scissions of the type described in chapter 3.

It is perhaps also surprising, in view of the fact that carbon–carbon bonds are significantly weaker than carbon–hydrogen bonds and that the effect of high-energy radiation is not expected to be specific, that carbon–hydrogen bond scission so clearly predominates over carbon–carbon bond scission. This may be due to the fact that the hydrogen atoms can diffuse relatively freely from the polymer side radical while main chain carbon–carbon bond scission would result in relatively immobile radicals which would not easily diffuse apart so that recombination would be highly probable.

Whether poly(methyl methacrylate) is subjected to high energy, UV radiation or to mechanical stress, the same nine-line ESR spectrum, shown in Fig. 7.5 is always obtained. It can be separated into a five-line spectrum with a 1 : 4 : 6 : 4 : 1 distribution of intensities and a four-line spectrum. There has been a sharp difference of opinion as to whether the two parts refer to the single radical XVI or whether this is accounted for only by the five-line

$$\underset{\text{XVI}}{\sim\!CH_2\!-\!\underset{\displaystyle \underset{COOCH_3}{|}}{\overset{\displaystyle \overset{CH_3}{|}}{C}}\cdot} \qquad \underset{\text{XVII}}{\sim\!CH_2\!-\!\underset{\displaystyle \underset{COOCH_3}{|}}{\overset{\displaystyle \overset{CH_3}{|}}{C}}\!-\!\dot{C}H\!-\!\underset{\displaystyle \underset{COOCH_3}{|}}{\overset{\displaystyle \overset{CH_3}{|}}{C}}\!-\!CH_2\!\sim} \qquad \underset{\text{XVIII}}{\sim\!CH_2\!-\!\underset{\displaystyle \overset{\displaystyle \overset{CH_3}{|}}{\dot{C}}}\!-\!CH_2\!\sim}$$

component, the remaining four lines being due to the chain side radical XVII. The balance of current opinion seems to favour the single radical XVI. High-energy irradiation results in chain scission to the exclusion of cross-linking and, as in photo-degradation (see chapter 3), the volatile products comprise carbon monoxide, carbon dioxide, methyl formate, methane and methanol, decomposition products of the radical $\cdot COOCH_3$, together with a little hydrogen. It seems that, as in photo-degradation, the initial scission occurs predominantly at the ester group forming radical XVIII and this is rapidly followed by chain scission to form radical XVI. If radical XVII was formed initially it would be even less likely than radical XVIII to combine in pairs to form cross-links because of the steric influence of the two substituents on each of the adjacent carbon atoms and the radical would be expected to undergo chain scission.

$$\text{XVII} \longrightarrow \quad \begin{matrix} & CH_3 & CH_3 & \\ & | & | & \\ \text{\textasciitilde\textasciitilde}CH_2-C-CH&=&C & + \cdot CH_2 \text{\textasciitilde\textasciitilde} \\ & | & | & \\ & COOCH_3 & COOCH_3 & \end{matrix} \qquad (7.16)$$

Materials like poly(α-methyl styrene) and polyisobutene, which, like poly(methyl methacrylate), have tetrasubstituted carbon atoms in the structural unit, undergo chain scission exclusively. A few polymers like polyethylene, poly(vinyl chloride) and poly(dimethylsiloxane) cross-link to

Figure 7.5. Electron spin resonance spectrum of poly(methyl methacrylate) irradiated and observed at room temperature.

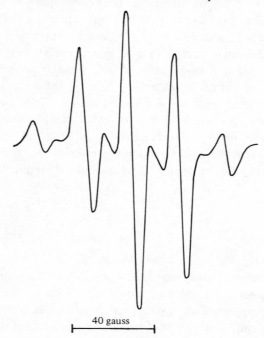

40 gauss

Table 7.1. *G values for chemical processes in irradiated polymers.*

Polymer	G(x)	G(s)	G (products)
polyethylene	3.5	—	4.0 (H_2); 2.4 (unsaturation)
polypropylene	0.9	0.6	2.5 (H_2); 0.1 (CH_4)
polystyrene	0.03	0.005	0.025 (H_2)
poly(vinyl acetate)	0.26	0.03	0.64 (H_2); 0.35 (CH_4); 0.21 (CO); 0.07 (CO_2)
poly(vinyl chloride)	2.15	—	(0.4 (H_2); 13 (HCl)
poly(dimethyl siloxane)	2.7	—	0.95 (H_2); 1.62 (CH_4); 0.22 (C_2H_6)
polyisobutene	–	4.1	1.4(H_2); 0.65 (CH_4)
poly(methyl methacrylate)	—	1.8	2.0 (total gas, see text)
poly(α-methyl styrene)	—	0.25	25 (α-methyl styrene); 0.035 (H_2); 0.035 (CH_4)
nylon 6,6	0.7	2.4	
poly(ethylene terephthalate)	2.3	0.07	0.25 (total gas; CO_2,CO,H_2,CH_4)

the virtual exclusion of scission but in the majority of polymers cross-linking and scission occur concurrently.

7.5.3. G values

The amount of energy absorbed by an irradiated material is usually measured in rads, one rad being defined as the absorption of 100 ergs/g. More relevant to the discussion of chemical mechanisms and yields, however, is the G value of radiation chemical yield which is the number of chemical events (cross-links, scissions, molecules of product, etc.) induced by 100 electron volts (eV) of energy. The G values for the reactions which occur in a number of common polymers are compared in table 7.1 in which G(s) and G(x) denote G values for chain scission and cross-linking respectively. It should be pointed out that there are big discrepancies in the literature but the values quoted in table 7.1 are representative. The commonest method of deriving G values for chain scission and cross-linking makes use of the Charlesby–Pinner equation which has already been referred to in chapter 3.

7.5.4. Radiation protection

It has been shown in chapter 5 that one effective method of stabilising polymer to UV radiation is by the use of UV-absorbing additives. This general method is ruled out for protection from high-energy radiation because the amount of energy absorbed is proportional to the number of electrons so that in presence of small concentrations of additives most of the

Table 7.2. *G (scission) values for*
copolymers of isobutene and styrene.

Styrene content (mole %)	G (scission)
0	5.9
20	3.0
50	1.8
80	1.0

energy will always be absorbed by the polymer. This implies that radical protection must depend upon interference by the protecting agent (antirad) with the reactions which occur subsequent to excitation of the electrons. Thus attention must be focussed upon the possibilities of transfer of the excitation energy before chemical reactions occur or upon scavenging of the radicals involved in these chemical reactions. Ion scavenging has been considered but found less effective.

Methods of protection have been described as internal or external, depending upon whether the protecting agent is built into the polymer structure or used as an additive. The data in table 7.1 demonstrate how polystyrene is relatively stable to high-energy radiation, almost certainly due to the fact that the aromatic rings act as an 'energy sink' in which the excitation energy is degraded to heat in their various vibration modes. Hence a degree of internal protection can be achieved by copolymerisation with styrene. Protection in this way is not spectacular but the data in table 7.2 for copolymers of isobutene and styrene are typical. It should be noted that the G values are based on the isobutene content and show that the presence of a styrene comonomer unit protects a number of isobutene units on either side of it.

A range of aromatic molecules have also been shown to act as low grade external protecting agents. Their action must be associated similarly with energy transfer and dissipation. For example, a variety of compounds like anthracene, biphenyl, benzoic acid, benzophenone and diphenylamine all give a measure of protection to PMMA. Other external protectives depend upon radical scavenging and it is not surprising that the hindered phenols and other types of anti-oxidants discussed in chapter 5 are effective. This effectiveness is much less than in thermal- and photo-oxidation and these protectives have the additional disadvantage that they tend to be themselves destroyed by the radiation.

All of these observations naturally lead to the conclusion that radiation stability will most likely be found in polymers with a high concentration of aromatic structures in the chain back-bone and the aromatic polyimides,

are probably the most effective to have been reported up to the present time.

7.6. Hydrolytic degradation

7.6.1. Polymer characterisation

Hydrolysis was the first topic in the field of polymer degradation to attract major attention when, in the 1920s and 1930s, it was applied as a tool in the investigation of the macromolecular structure of polysaccharides. It was shown that the hydrolysis of the glucosidic linkage occurs more or less at random, resulting in chain scission, a rapid decrease in molecular weight and the ultimate production of high yields of the monomer, glucose. Later, the controlled hydrolysis of proteins and nucleic acids followed by the identification of products played a vital part in the elucidation of the structures of these complex substances.

7.6.2. Polymer deterioration

Hydrolysis is an important deteriorative process in modern polymer technology because many commercial materials incorporate hydrolysable linkages. These include polyesters, polyamides, polyurethanes, poly(dialkyl siloxanes) and polycarbonates. In each of these groups, except some polyesters, in which the ester group is pendant to the chain, the hydrolysable group is almost invariably part of the main chain structure so that hydrolysis results in chain scission. This implies a rapid decrease in molecular weight and consequent rapid deterioration in mechanical properties such as tensile strength. Indeed the extent of reaction may be measured quantitatively in terms of chain scissions using formula (*iii*) in chapter 2 or by analysis of the new end-groups formed, or qualitatively, although often important from the practical point of view, from measurements of tensile strength.

In polyesters, on the other hand, the ester groups may be present either in the main chain, as in poly(ethylene terephthalate) and the polyester polyurethanes or as pendant groups as in the polymethacrylates and poly vinyl esters. When the ester groups are pendant, chain scission does not occur on hydrolysis and the physical properties are not affected by small proportions of modified pendant groups.

Hydrolysis of polyacrylonitrile can affect physical properties because the amide and carboxyl groups resulting from nitrile hydrolysis, reaction (7.17)

$$\cdots CH_2-CH\cdots \xrightarrow{H_2O} \cdots CH_2-CH\cdots \longrightarrow \cdots CH_2-CH\cdots$$
$$| \qquad\qquad\qquad | \qquad\qquad\qquad |$$
$$CN \qquad\qquad\qquad C \qquad\qquad\qquad C$$
$$O^{\nearrow\ }{}^{\searrow}NH_2 \qquad\qquad O^{\nearrow\ }{}^{\searrow}OH \qquad (7.17)$$

can act as initiating centres for subsequent thermal degradation as described in chapter 2. On the other hand, surface hydrolysis of poly-acrylonitrile films may be used to produce hydrophilic groups for the attachment of dye molecules.

Polymers are generally much more resistant to hydrolysis than might be predicted from the behaviour of low molecular weight model compounds. This is not because of any inherent difference in the stability of the hydrolysable linkages but because attack is usually confined to the surface because of the hydrophobic nature of most organic polymers and the consequent low rate of diffusion of aqueous solutions into the polymer bulk. Crystalline polymers are also usually less susceptible because of the inaccessibility of the crystalline as compared with the amorphous regions. Traces of moisture from the preparation of certain condensation polymers may, of course, be trapped within the mass of the polymer so that hydrolysis may compete with thermal degradation at higher temperatures. There is evidence that the thermal degradations of poly(ethylene terephthalate) and the polyamides are influenced in this way. There is also evidence that the acid groups formed in the hydrolysis of polyesters may catalyse further reaction thus conferring autocatalytic properties on the overall process.

Commercial polyurethanes are prepared by the condensation of diisocyanates with polyester or polyether diols.

OCN------NCO + HO-polyester or polyether-OH

$$\downarrow$$

$$\begin{matrix} O & H & & H & O & & & & O & H & & H & O \\ \| & | & & | & \| & & & & \| & | & & | & \| \\ O-C-N-----N-C-O\text{-polyester or polyether-O} & -C-N-----N-C-O \end{matrix}$$

$$(7.18)$$

Although the choice between polyester and polyether for any given application is governed predominantly by the mechanical properties required in the polymer, ether based polyurethanes are preferred when laundering in hot water with alkaline detergents because of the greater susceptibility of polyesters to hydrolytic attack. On the other hand, polyesters are more resistant to organic solvents and are preferred to polyethers when the polymer is to be subjected to dry cleaning. The relative susceptibility of ester and ether-based polyurethanes may be gauged from the fact that a polyester polyurethane has been shown to lose 90% of its tensile strength (due to hydrolytic chain scission) after 18 months at 25 °C and 100% relative humidity whereas a polyether polyurethane was unaffected under the same conditions.

In the overall weathering process in many polymers there is no doubt that humidity and the presence of liquid water often play a significant role, but it is usually difficult to disentangle the effects of the large number of degradation agencies at work in typical outdoor exposures where the effects of sunlight and oxygen are clearly paramount. There are some situations, however, in which there is no doubt that hydrolysis is the most important agency and these have usually involved poly(ethylene terephthalate) or polyamides. The tensile properties of fibres forming polymers are extremely sensitive to reduction of molecular weight by chain scission. For example, polyesters have been shown to be unsuitable in certain underground applications, especially those in which they came in contact with acid clays.

7.6.3. Recycling of polymers by hydrolysis

As a result of the continuously rising cost of petroleum feedstocks, from which most commercial polymers are derived, as well as the promise of progressively more and more severe shortages in the foreseeable future, attention is being given increasingly to the recycling of polymers. The term 'recycling' is used to describe at least three aspects of the conservation of the raw materials of polymer preparation.

First, it envisages that scrap polymer may be collected and burned to produce energy. Since most high-tonnage commercial polymers, with the obvious exception of poly(vinyl chloride), are easily flammable this should be relatively easy because it would require only the minimum of separation of the various plastic materials. It is not at present very attractive economically, however, due to the capital cost involved.

Secondly, plastic scrap may be collected for reprocessing. In this case, apart from the problems of collection, segregation of the various types of materials is a major problem. In addition, aged polymer may contain the seeds, especially in the form of peroxides, of its own rapid destruction during the second round of application so that stabilisation problems will tend to become more acute.

Finally it may be possible, by suitable treatment, to recover useful chemicals from the scrap polymer and it is to this aspect that most attention has been given to date.

By the very chemical nature of step-growth polymers like the polyesters, polyamides, polyurethanes and polycarbonates, hydrolysis should be expected to give high yields of the monomers or of compounds closely related to the monomers. Since, in many step-growth polymers, these monomers, by reason of their chemical nature, represent a larger fraction of the cost of production than the monomers of addition polymers, this approach to polymer recycling is currently the most attractive economically. Thus diamines may be recovered from polyamides, terephthalic acid and ethylene glycol from poly(ethylene terephthalate) and glycols and amines from polyurethanes.

It would seem that the scrap car industry, which currently exploits only the scrap metal, represents a considerable volume of easily segregated scrap plastic which could form the basis for polymer recycling of this kind.

Suggested further reading

Degradation in polluted atmospheres

1. H. H. G. Jellinek, in *Aspects of Degradation and Stabilisation of Polymers*, ed. H. H. G. Jellinek, Elsevier 1978.
2. R. M. Harrison and C. D. Holman, Ozone Pollution in Britain, *Chem. in Brit.*, **18**, 563, 1982.
3. S. D. Razumovskii and C. E. Zaikov, Degradation and Protection of Polymeric Materials in Ozone, *Developments in Polymer Stabilisation – 6*, ed. G. Scott, App. Sci. Pub., London, 1982.

Degradation at high temperatures

1. C. David, in Comprehensive Chemical Kinetics, Vol. 14, *Degradation of Polymers*, ed. C. H. Bamford and C. F. H. Tipper, Elsevier, 1975.
2. W. W. Wright, in *Degradation and Stabilisation of Polymers*, ed. G. Geuskens, App. Sci. Pub., London, 1975.
3. E. L. Strauss, in *Aspects of Degradation and Stabilisation of Polymers*, ed. H. H. G. Jellinek, Elsevier, 1978.

Mechanical and ultrasonic degradation

1. A. M. Basedow and K. H. Ebert, *Adv. Polym. Sci.*, **22**, 83, 1977.
2. K. Murakami, in *Aspects of Degradation and Stabilisation of Polymers*, ed. H. H. G. Jellinek, Elsevier, 1978.
3. N. K. Baraboim, *Mechanochemistry of Polymers* (trans. R. J. Moseley, ed. W. F. Watson), Unwin, 1964.
4. J. Sohma, in *Developments in Polymer Degradation – 2*, ed. N. Grassie, App. Sci. Pub., 1979.
5. W. Schnabel, *Polymer Degradation*, Hanser, 1981.

Degradation by high-energy radiation

1. C. David, in *Degradation of Polymers*, eds. C. H. Bamford and C. F. M. Tipper, Elsevier, 1975.
2. B. J. Lyons and V. L. Lanza, in *Polymer Stabilisation*, ed. W. H. Hawkins, Interscience, 1972.
3. W. Schnabel, in *Aspects of Degradation and Stabilisation of Polymers*, H. H. G. Jellinek, Elsevier, 1978.
4. A. Charlesby, in *Polymer Science, a Materials Science Handbook*, ed. A. D. Jenkins, North Holland, 1972.

Hydrolytic degradation

1. D. A. S. Ravens and J. E. Sisley, in *Chemical Reactions of Polymer*, ed. E. M. Fettes, Interscience, 1964.
2. B. D. Gesner, in *Polymer Stabilisation*, ed. W. L. Hawkins, Interscience, 1972.

INDEX

ablation, 199–201
abnormal structures, 18
ABS – *see* copolymers of acrylonitrile, butadiene *and* styrene, 16
accelerated ageing tests, 10
acoustic energy, 202
additive fire retardants, 173
afterglow, 188
ageing, 1, 68, 80
aggressive environments, 190–216
4-alkylaminodiphenylamines, 153
alum, 187
aluminium oxide, 187
β-amino-crotonate esters, 146
ammonium polyphosphate as fire retardant, 177
antagonism, 122, 162, 164
antifatigue agents, 119, 154
antimony oxide, 183, 185–7
antioxidants, 4, 97, 119–69
antioxidant 2246, 166
antioxidants for rubbers, 123
antioxidants, non-staining, 123
antioxidants, peroxidolytic, 127–33
antiozonants, 119, 154
antirad, 212
aromatic nylon, 197
atmospheric pollutants, 3, 115
auto-acceleration of oxidation, 89
auto-retardation, 110
autoxidation, 2
azobisisobutyronitrile, 185

benzimidazoles, 198
benzoxazoles, 198
benzoylacetone, 146
benzthiazyl sulphenamides, 153
biodegradation, 4, 6
biphenyl ethers as fire retardant additives, 184
blends of poly(vinyl chloride) and poly(methyl methacrylate), 63
blends of poly(vinyl chloride) and poly(vinyl acetate), 65
borax as fire retardant, 188

boric acid as fire retardant, 188
boron compounds as fire retardants, 188–9
brominated aliphatic ethers as fire retardant additives, 184
brominated biphenyls as fire retardant additives, 184
bromine compounds as fire retardants, 182–5
2-bromoethyl methacrylate, 184
butyl rubber, 191

cage recombination, 75, 117
candle test, 171–2
capillary rheometer, 97
carbon black, 135, 158
carbon fibres, 49, 52–3, 200
carbonyl index, 16
casein, 204
cavitation, 203
cellulose, 176
chain branches, 18
chain breaking antioxidants, 120–7
chain scission, 69, 70, 73, 78, 82, 84, 95, 96, 99, 109, 113, 191–2, 208
Charlesby–Pinner equation, 73, 211
chelating agents, 133
chlorinated alkanes as fire retardants, 182
chlorinated rubber, 20
chlorine compounds as fire retardants, 182–5
chlorophyll, 116
chromatographic methods of anlysis, 14
cold ring fraction, 23
copolymers of, acrylonitrile, butadiene and styrene, 16, 93, 156; acrylonitrile and 2-bromoethyl methacrylate, 19, 184; ethylene and carbon monoxide, 69; ethylene and propylene, 54; methyl acrylate and 2-bromoethyl methacrylate, 184; methyl methacrylate and alkyl acrylates, 59; methyl methacrylate and 2-bromoethyl methacrylate, 173–4, 184, methyl methacrylate and α-chloroacrylonitrile, 55; methyl methacrylate and methyl acrylate, 73–6,

218